JOURNAL OF APPLIED LOGICS - IFCOLOG
JOURNAL OF LOGICS AND THEIR APPLICATIONS

Volume 8, Number 9

October 2021

Disclaimer

Statements of fact and opinion in the articles in Journal of Applied Logics - IfCoLog Journal of Logics and their Applications (JALs-FLAP) are those of the respective authors and contributors and not of the JALs-FLAP. Neither College Publications nor the JALs-FLAP make any representation, express or implied, in respect of the accuracy of the material in this journal and cannot accept any legal responsibility or liability for any errors or omissions that may be made. The reader should make his/her own evaluation as to the appropriateness or otherwise of any experimental technique described.

ISBN 978-1-84890-378-4
ISSN (E) 2631-9829
ISSN (P) 2631-9810

College Publications
Scientific Director: Dov Gabbay
Managing Director: Jane Spurr

http://www.collegepublications.co.uk

iii

SCOPE AND SUBMISSIONS

This journal considers submission in all areas of pure and applied logic, including:

pure logical systems
proof theory
constructive logic
categorical logic
modal and temporal logic
model theory
recursion theory
type theory
nominal theory
nonclassical logics
nonmonotonic logic
numerical and uncertainty reasoning
logic and AI
foundations of logic programming
belief change/revision
systems of knowledge and belief
logics and semantics of programming
specification and verification
agent theory
databases

dynamic logic
quantum logic
algebraic logic
logic and cognition
probabilistic logic
logic and networks
neuro-logical systems
complexity
argumentation theory
logic and computation
logic and language
logic engineering
knowledge-based systems
automated reasoning
knowledge representation
logic in hardware and VLSI
natural language
concurrent computation
planning

This journal will also consider papers on the application of logic in other subject areas: philosophy, cognitive science, physics etc. provided they have some formal content.

Submissions should be sent to Jane Spurr (jane@janespurr.net) as a pdf file, preferably compiled in LaTeX using the IFCoLog class file.

CONTENTS

ARTICLES

PREFACE

MICHAŁ ARASZKIEWICZ
Jagiellonian University, Kraków
`michal.araszkiewicz@uj.edu.pl`

TOMASZ ZUREK
Maria-Curie Sklodowska University, Lublin
`tomasz.zurek@mail.umcs.pl`

Reasoning with and about evidence is arguably the most complex domain of legal reasoning (cf. [7]). The aim of all evidentiary proceedings is to determine the factual basis for a legal decision. Arguably, then, the aim of reasoning with and about legal evidence is to obtain the set of sentences that adequately represent what happened in the case. However, this view of evidentiary proceedings may be overly simplistic, and for several reasons, among which the following should be underlined: (1) the connections and dependencies between legal evidentiary reasoning and the legal regulation; (2) the interdisciplinary character of knowledge relevant for reasoning with and about legal evidence and (3) the problems of structure, validity and justification of the evidentiary reasoning itself.

Consequently, first, the determination of the factual basis of the case always takes place in a certain legal setting. The rules and principles of law both foster, guide and constrain the process of reasoning of the different actors involved in the proceedings. In certain cases, the applicable principles obligate the relevant actors to attempt to provide the true picture of the case while in other settings it is sufficient to obtain the probable or plausible version. Certain legal contexts require the acceptance of a fictitious or even counterfactual version of the event, for example supported by the applicable legal presumptions and no satisfactory proof has been provided to the contrary. In some jurisdictions, the notion of standard of proof plays a central role in the process of fact finding; hence, the facts have to be proven beyond reasonable doubt or on the basis of preponderance of evidence or of another applicable standard. In some legal systems, the assessment process performed by a fact finder may be subject to a detailed regulation or may be left to evaluation based on plain reason, "rules of logic and experience" or other similar open standards. Eventually, the law also determines the types of applicable sources of proof and regulates the procedure of fact finding, requiring the decision maker to make certain decisions on the basis of legal presumptions and rules concerning the distribution of burden of proof. As is commonly known, the very notion of burden of proof is ambiguous (e.g. [34]: 1339–1415), subject to different concretisations in different jurisdictions, which poses a significant challenge for logical and computational modelling ([10]; [19]; [6]). Therefore, reasoning with and about legal evidence requires a significant degree of legal expertise.

Second, legal fact finding is a highly interdisciplinary domain of investigation, and the degree of its interdisciplinary character increases at a fast pace. This is evident particularly when establishing the facts of the case requires evidence based on expert witness opinions. The procedural institution of expert witness opinions is based on the structure of the argument from expert opinion, one of the most thoroughly investigated types of argument ([27]). On a day-to-day basis, the courts are presented with and are obligated to assess the opinions of expert physicians, biologists, chemists, information technology (IT) experts, engineers of different specialties and others. In this area, the sphere of legal reasoning overlaps with scientific reasoning in the sense that the scientific findings provide the premises for the determination of the factual state of affairs. The omnipresence of evidence based on expert knowledge in legal fact finding has given rise to new and extensively debated problems (e.g. [33]). How can a trier of fact assess the expert opinion when she or he does not have expert knowledge in such field? However, if this task is complicated, does this not mean that the administration of justice is actually placed in the hands of the experts in other fields ([1])? How should the legal decision maker decide in the face of disagreement among the experts? How should the methodological problems within the given branch of science itself be taken into account in the area of law? In the first place, the requirement of grounding the legal fact finding on the current scientific state of the art is less evident than it seems when one is aware of the problems related to the scientific method. On the other hand, other than science, there is no area of discourse that the law can make recourse to for painting the true picture of reality. Therefore, in spite of the numerous problems concerning the role of scientific reasoning in the domain of legal fact finding, there is no viable alternative to integrating the two areas. The increasing awareness of the methodological problems concerning the establishment of scientific knowledge does not alter the conclusion that falsifiable and rigorous scientific results provide the best available approximation of the accurate picture of reality and the nature of the human mind. It should also be emphasised that technological progress, being a natural consequence of the development of the natural sciences, brings forth new possibilities but also risks with regard to legal fact finding. In particular, the development of IT, including the solutions collectively referred to as artificial intelligence

(AI), on the one hand enables the analysis of big datasets and yields results that used to be unavailable. On the other hand, though, the use of these technologies gives rise to new concerns related to the reliability of data (which in turn leads to the broad problems of cybersecurity), the ethical and legal concerns connected to the problems of bias in data and to the explainability of algorithms and the problems concerning the normative relevance and justification of the results of intelligent solutions, among others (see on various aspects of these problems [13]; [13]; [21]).

Third, it is necessary to keep in mind that evidentiary decision making (and reasoning) is a subcategory of human decision making (especially reasoning), and as such, is prone to error and bias. The research in cognitive psychology for more than five decades has revealed the multifarious limits of human rationality, particularly the systematic biases concerning the assessment of probability, the framing effect, the impact of different types of heuristics, and others ([11]). The cognitive-psychology research has also highlighted different problems concerning human memory and perception, which may also lead to erroneous conclusions. Moreover, the triers of fact (and the sources of evidence, such as witnesses and experts) may be subject to political bias, stereotypical thinking or manipulative persuasion. Psychological knowledge may mitigate the negative effects of the aforementioned phenomena, but again, it should be taken into account that cognitive psychology is itself a heterogeneous, quickly evolving field where many of the competing theories and models are incompatible.

One should note, however, that while decision making and reasoning are psychological processes, in the context of law, a rational justification of the decisions made should be provided and appropriate legal arguments supporting the conclusions should be presented. Therefore, at the outset, the legal evidentiary reasoning should be valid with regard to both its structure and the premises adopted. The nature of the standards of this validity is subject to debate while reasoning with and about legal evidence does not have a strictly deductive structure in most cases; that is, in most instances, it does not involve reasoning from a well-established set of true premises to a conclusion. Legal fact finding is a classic example of a domain where reasoning with imperfect (incomplete, inconsistent, uncertain) information is necessary, which leads to specific challenges, such as how to fill the knowledge gaps of the legal fact finder and how to justify the choices related thereto, how to resolve the conflicts between inconsistent pieces of evidence such as contradictory witness testimonies and how to handle the uncertainty of the judgment taking into account the advances of the theories of probabilistic inference. Considering the significance and depth of these challenges, it comes as no surprise that legal reasoning with facts has been the subject of systematic investigations in different theories of reasoning for almost a hundred years now.

Among the pioneering works on the foregoing, the famous conception of the Wigmore diagram ([36]), whose author intended to develop a scientific fact-finding method with legal evidence, should be mentioned. Since the emergence of such conception, numerous models of evidential reasoning have been developed in interdisciplinary settings. These models attempt to represent and emphasise the rational elements of evidence-related reasoning and to indicate the possible limits of the rationalisation of the fact-finding process, identifying the spheres of potentially arbitrary or erroneous decision making. Generally speaking, the three dominant approaches at present the probabilistic approach; the scenario-, narrative-, or story-based approach and the argumentation-based approach ([7]; [18]). The probabilistic approach applies the tools and concepts of the mathematical theory of probability to legal fact finding. Among the many approaches, Bayesian network modelling has gained prominence due to its widespread applications in the field of AI (for a recent overview of this approach, see [23]). The story-based approach represents the circumstances of cases in terms of complex sets of elements, which have to satisfy certain criteria, particularly the conditions of coherence. A good theory of the facts of the case should also explain the causal relations among the events ([4]; [5]). Finally, the argumentation-based approach focuses on the comparison of the relative strengths of the reasons for and against the adoption of a certain conclusion. This approach is particularly internally differentiated as it has very distinct theories and models, including the informal topical-rhetorical approach ([17]); the theories of argumentation based on the notion of rational communication, such as pragma-dialectics ([8]); the argumentation schemes approach ([31]) and the various other models developed in the field of computational argumentation ([20]; [3]). The argumentation-based models of reasoning with and about the theories of evidence have been developed and investigated in the research area of AI and law ([29]), but it must be stressed that in recent years, these three aforementioned approaches have been combined to overcome the limitations of particular perspectives and to bring about new scientific results particularly in the field of AI and law. The aforementioned contribution by [4] is a hybrid theory combining the story- and argumentation-based approaches. Among the many other contributions are the use of argumentation tools to discuss the Bayesian analyses of criminal cases ([18]), the use of Bayesian networks to compare the strengths of legal arguments ([15]), the construction of Bayesian network graphs from legal arguments ([35]) and the extraction of explanatory arguments from Bayesian networks ([22]). Further hybridisation of the models developed in the literature of the subject may be expected in the future. The deepening of the interdisciplinary perspective in the investigation of legal evidentiary reasoning brings forth important new insights but at the same time leads to new open questions.

Taking into account the three contexts of the research on reasoning with and about legal evidence (i.e. the context of legal regulation, the context of multi- and interdisciplinary research relevant for the problems of evidence and the context of theories and models of legal reasoning), let us now rephrase the problems of this domain of legal reasoning in the following manner. Each theoretical account of legal evidentiary evidence should consider the layers cited below.

The layer of available data (what can be known at the outset about the facts of the current situation) Here, the aforementioned development of new theoretical and technological tools already plays a pivotal role, and this role is expected to increase in the future. The layer of reasoning schemes and patterns aiming at establishing the facts of the case, including the probabilistic, story-based and argumentative approaches discussed earlier and the specific patterns and constraints accepted in the legal culture under consideration Here, it seems necessary to stress that the relevant reasoning may be modelled with the use of any of the approaches outlined earlier, and specifically with the use of hybrid models. The normative layer, which is directly interested in the practical question of what should be done, which includes legal-political considerations, balancing of values, the constraints arising from the existing legal regulation and the general underlying assumptions concerning the rationality of legal fact finding and their relation to the persuasiveness of argumentation It should be noted that the conclusions on this third layer may be to some extent independent from or even contrary to the conclusion on the preceding layer. The special issue of the *IfCoLoG Journal of Logics and Their Applications* devoted to the article "Methodologies for Research on Legal Argumentation: Argumentation and Evidence" aimed to explore the selected trends and topics in the current state of the art in developing methods and conceptual frameworks in the study of legal evidential argumentation. The main objective of the issue was to provide space for the presentation of methodological ideas concerning the research on legal argumentation in the context of evidential reasoning.

The issue was founded on the discussions that took place at the 2nd MET-ARG: "The International Workshop for Methodologies for Research on Legal Argumentation" held on September 15, 2018 in Warsaw under the auspices of the ArgDiaP organisation in conjunction with WAW (Warsaw Argumentation Week) 2018.

As for the present special issue of the *IfCoLoG Journal of Logics and Their Applications*, it brings together a selection of insightful papers that address a wide range of topics related to the methods and methodology of research on legal argumentation in the area of evidential reasoning. The special issue opens with the paper authored by Federico Costantini, Fausto Galvan, Marko Alvise De Stefani and Sebastiano Battiato entitled "Assessing 'Information Quality' in IoT Forensics: Theoretical Framework and Model Implementation", in which the authors explore the problem of the Internet of Things (IoT) forensics, especially the issue of data and information quality in IoT in the quickly evolving branch of forensic science, referred to as digital forensics. The authors introduce a theoretical framework on data quality and information quality, where they focus on forensic-analysis challenges in IoT environments, providing a use case of evidence collection for investigative purposes. At the end of the article, the authors propose a formal framework for assessing the information quality of IoT devices for forensics analysis. The key point of the paper is the distinction of three perspectives on information and information quality: (1) information as reality; (2) information about reality; and (3) information for reality. For each of these three perspectives, the authors introduce a formal model of quality of information (QoI) measure along with formulae for the measure's calculation. The results were verified on the basis of two real-life case studies. The paper offers a perspective on the first of the three layers of evidential reasoning described earlier: the layer of data and information available through IT. It is also an example of the phenomenon of hybridisation of the models used to tackle legal evidential problems, providing a set of mathematical equations whose application may provide the point of departure for evidential argumentation.

The second paper, authored by Anne Ruth Mackor and entitled "Different Ways of Being Naked: A Scenario Approach to the Naked Statistical Evidence Problem", contributes to the ongoing dispute concerning the relevance of the so-called naked statistical evidence in legal reasoning about facts. As is well known, statistical inference often brings about results that imply the conclusions about the factual layer of the cases, particularly because they seem to satisfy the requirements of the applicable standards of proof. However, for many scholars, the results brought forth by this type of reasoning seem insufficient, although it is generally difficult to explain why this is the case; it is commonplace, however, to demand that the evidence, at least in a criminal case, be "individuated" or "specific". Mackor discusses this problem by introducing the scenario-based approach ([16]; [24]), which allows a more robust modelling of reasoning with legal evidence.It is emphasised in the literature that the scenario-based approach provides a more solid justification for the theses about the states of affairs because it introduces the layer of causality, which makes the evidence more individuated or specific; thus, it goes beyond statistical plausibility. Moreover, the scenario-based approach encompasses the comparative evaluation of the competing explanations of the evidence and the hypothetical reasoning: the explanation of evidence that may have been caused by other states of affairs, actions or events hypothesised by one or more of the scenarios about the given factual situation. Mackor further extends the employed conceptual scheme by applying the distinction between trace evidence (evidence that may have been brought forth by the state of affairs or by an event included in the scenario) and predictive evidence (information relevant for the evaluation of the scenario's plausibility but which could not have been caused by the persons or events involved in the scenario). Then the author uses further, more fine-grained distinctions of trace and predictive evidence to classify the different types of naked statistical evidence present in the discussed cases. The paper itself does not argue directly for or against the use of naked statistical evidence in legal trial, but proposes a systematisation of the evidence commonly classified under this label. The paper may be situated in the second of the three layers earlier distinguished: the layer of interplay between the different patterns of reasoning used to establish a plausible version of the event in legal evidentiary reasoning.

Finally, the third paper, authored by Douglas Walton (+2020) and entitled "Using Distance in Argument Maps to Model Conditional Probative Relevance", belongs to the sphere of formal and computational modelling of practical argumentation. The paper offers a perspective on the modelling of the crucial but notoriously opaque notion of legal relevance, taking the Federal Rules of Evidence as an illustrative material. The paper argues that the concept of relevance may be explicated through the use of the notion of inferential distance proposed in an earlier paper by [12] and modelled with the use of graphs in the style of the Carneades Argumentation System ([9]; [10]). The author distinguishes the probative relevance of sentences from the topical relevance of sentences. Two sentences are probatively relevant to each other if one of them may be used to argue for or against the other. Topically relevant sentences, on the other hand, have the same subject matter but may be nonetheless not probatively relevant to each other. The paper analyses the notion of probative relevance and formalises it with the use of the conceptual framework of the Carneades Argumentation System. The paper also discusses the concept of conditional probative relevance: the relation of probative relevance that would obtain if a certain sentence is supported by the evidence of the case. The paper by Douglas Walton is an excellent example of a contribution applying a formal approach to explicate the notion employed in legal regulation. The approach proposed there may also be referred to as a hybrid one as it combines such tools as argument maps and argumentation schemes.

The final paper featured in this special issue may be the last published work of Douglas Walton. Thus, this is a good occasion to emphasise the influence of his work on the understanding of the problems of reasoning with and about legal evidence.

Evidential reasoning in law and in other areas was one of the key points of Douglas Walton's research. He authored a number of books strictly addressing this issue ([28]; [30]; [32] and many others) and countless journal and conference papers. He was best known for his research on argumentation schemes and critical questions ([25]; [26]; [31] and many others), which became a background for many of the aforementioned argumentation-based evidential reasoning models and hybrid argumentation- and story-based models, in which many argumentation schemes can be seen as evidential generalisations while critical questions can be used to examine evidential reasoning. Moreover, the dialogical view on argumentation (which was also one of the central points of his research) allowed for the modelling of reasoning about the burden of proof and other proof standards with the use of defeasible argumentation schemes and critical questions. In general, his work was a genuine bridge between informal and formal models of legal reasoning, allowing the gap between these two approaches to be filled. As far as the specific problems are concerned, Douglas Walton, being one of the leading researchers on the problem of argument based on expert opinion, applied the formal models to analyse argumentation based on expert witness opinion in law ([33]). The topic of Walton's contribution to the reasoning theory in general and to the theory of reasoning with legal evidence in particular requires a thorough interdisciplinary investigation. Walton's influence on AI and law has recently been discussed in [2], and an issue of this journal has been devoted to a broader perspective on his work [37]. These publications offer a good starting point for broader and deeper studies on Walton's work and for the future research based on or inspired by his legacy.

We, the guest editors, would like to express our appreciation of the Program Committee's substantial efforts to review the papers submitted for inclusion in this special issue. Likewise, we thank all the authors for submitting their excellent manuscripts for this special issue. We also express our sincere thanks to the Editorial Team of the IfCoLoG Journal of Logics and Their Applications.

References

[1] Allen, Ronald Jay, Fiddling While Rome Burns: The Story of the Federal Rules and Experts (October 1, 2017). *Symposium on Forensic Expert Testimony*, Daubert and Rule 702 Northwestern Public Law Research Paper No. 17-29, 2017. (Available at SSRN: https://ssrn.com/abstract=3080628)

[2] Atkinson, K. and Bench-Capon, T. and Bex, F. and Gordon, T. F. and Prakken, H. and Sartor, G. and Verheij, B.: *In memoriam Douglas N. Walton: the influence of Doug Walton on AI and law*. Artificial Intelligence and Law, 28(3), 281-326, 2020

[3] Baroni P., Gabbay D., Parent X., van der Torre L. (eds.) *Handbook of Formal Argumentation*, College Publications 2018

[4] Bex F.: *Arguments, Stories and Criminal Evidence. A Formal Hybrid Theory*, Springer, 2011

[5] Bex F., Verheij B. Legal Stories and the Process of Proof. *Artificial Intelligence and Law* 21 (3), 253-278, 2013

[6] Calegari R, Sartor G: A Model for the Burden of Persuasion in Argumentation. *Proceedings of 33rd International Conference on Legal Knowledge and Information JURIX 2020*, pp.: 13-22, IOS Press 2020

[7] DiBello M., Verheij B.: Evidential Reasoning. *Handbook of Legal Reasoning and Argumentation* eds. Bongiovanni, G., Postema, G., Rotolo, A., Sartor, G., Valentini, C., and Walton, D., pp.: 447-493, Springer 2018

[8] Eemeren F.H. v, Grootendorst R.: *Argumentation, communication, and fallacies: A pragma-dialectical perspective*, Lawrence Erlbaum Associates, 1992

[9] Gordon, T. F. and Walton, D. The Carneades argumentation framework âĂŤ using presumptions and exceptions to model critical questions. *Proceedings of the First International Conference on Computational Models of Argument (COMMA 06)*, IOS Press, 2006

[10] Gordon, T.F. and Prakken, H. and Walton, D.N.: The Carneades model of argument and burden of proof. *Artificial Intelligence* 171: 875-896, 2007.

[11] Kahneman D.: *Thinking, Fast and Slow*, Penguin Books, 2011

[12] Macagno F.: Assessing Relevance, *Lingua*, July–August, 2018, 42-64, 2018.

[13] Nalepa, G. and Araszkiewicz, M. and Nowaczyk, S. and Bobek, S.: Building trust to AI systems through explainability: technical and legal perspectives, in Nalepa, G. and Atzmueller, M. and Araszkiewicz, M. and Novais, P. (eds) *XAILA 2019 EXplainable AI in Law 2019: proceedings of the 2nd EXplainable AI in Law Workshop (XAILA 2019)*, Madrid, Spain, December 11, 2019

[14] Nalepa, G., Araszkiewicz, M.: Liability and responsibility of intelligent systems. In GonzÃąlez-Espejo MarÃŋa JesÃžs, PavÃşn Juan (eds.): *An introductory guide to artificial intelligence for legal professionals*, Wolters Kluwer, 2020

[15] Neil M., Fenton N., Lagnado D., Gill R.D.: Modelling competing legal arguments using Bayesian model comparison and averaging, *Artificial Intelligence and Law* 27, 403-430, 2019

[16] Pennington, N., and Hastie, R., The story model for juror decision making. In R. Hastie (ed.), *Inside the jury: The psychology of juror decision making*, Cambridge University Press, 1993 (2nd ed.), pp. 192-221.

[17] Perelman Ch., Olbrechts-Tyteca L.: *The New Rhetoric: A Treatise on Argumentation*, University of Notre Dame Press, 1973

[18] Prakken H., Bex F., Mackor A. R: Editors' Review and Introduction: Models of Rational Proof in Criminal Law, *Topics in Cognitive Science* Vol. 12(4), 1053-1067, 2020

[19] Prakken, H. and Sartor, G.: A logical analysis of burdens of proof. In H. Kaptein, H. Prakken and B. Verheij (eds.) *Legal Evidence and Proof: Statistics, Stories, Logic*, Ashgate Publishing, Applied Legal Philosophy Series, 2009, pp. 223-253.

[20] Rahwan I., Simari G. (eds.): *Argumentation in Artificial Intelligence*, Springer, 2009

[21] Sovrano, F. and Vitali, F. and Palmirani, M.: The Difference between Explainable and Explaining: Requirements and Challenges under the GDPR. In Nalepa, G., Atzmueller, M., Araszkiewicz, M., and Novais, P. (eds) *XAILA 2019 EXplainable AI in Law 2019:*

proceedings of the 2nd EXplainable AI in Law Workshop (XAILA 2019), Madrid, Spain, December 11, 2019

[22] Timmer, Sjoerd T. and Meyer, John-Jules Ch. and Prakken, H. and Renooij, Silija and Verheij, Bart: A two-phase method for extracting explanatory arguments from Bayesian networks. *Int. J. Approx. Reason.* 80: 475-494, 2017

[23] Urbaniak R., Di Bello M.: Legal Probabilism, *The Stanford Encyclopedia of Philosophy* (Summer 2021 Edition), Edward N. Zalta (ed.), URL = https://plato.stanford.edu/archives/sum2021/entries/legal-probabilism/, 2021.

[24] Wagenaar, W. A., Van Koppen, P. J., and Crombag, H. F. M. *Anchored narratives: The psychology of criminal evidence.* London: Harvester Wheatsheaf, 1993

[25] Walton, D. and Krabbe, E. *Commitment in dialogue: basic concepts of interpersonal reasoning.* SUNY Press, Albany, 1995

[26] Walton, D.: *Argumentation schemes for presumptive reasoning.* Lawrence Erlbaum Associates, Mahwah, 1996

[27] Walton D.: *Appeals to Expert Opinion: Arguments from Authority*, Pennsylvania State University Press, 1997

[28] Walton, Douglas: *Legal Argumentation and Evidence.* Pennsylvania State University Press, 2002.

[29] Walton D.: *Argumentation methods for artificial intelligence in law.* Springer, 2005

[30] Walton, Douglas: *Witness Testimony Evidence: Argumentation and the Law.* Cambridge University Press, 2007.

[31] Walton D., Reed C., Macagno F.:*Argumentation Schemes*, Cambridge University Press, 2008

[32] Walton Douglas. *Argument Evaluation and Evidence.* Springer Publishing Company, 2016.

[33] Walton, D.: *When expert opinion evidence goes wrong.* Artif Intell Law 27, 369–401, 2019.

[34] Weinstein J.B., Abrams N., Brever S., Medwed D.S.: *Evidence. Cases and Materials*, 10th ed., Foundation Press

[35] Wieten, G. M. and Bex, F. J. and Prakken, H., and Renooij, S.: Exploiting causality in constructing Bayesian network graphs from legal arguments. *Proceedings of Jurix 2018*, pp. 151-160. IOS Press. https://doi.org/10.3233/978-1-61499-935-5-151 2018

[36] Wigmore F.: The Principles of Judicial Proof, Little, Brown & Company, 1931

[37] Woods, J. (eds): *Special issue: Douglas Walton Remembered.* Journal of Applied Logics - IfCoLog Journal, vol. 8, no. 1, 2020

Received July 2021

Assessing "Information Quality" in IoT Forensics: Theoretical Framework and Model Implementation*

Federico Costantini
Researcher at the Department of Law
University of Udine (IT)
federico.costantini@uniud.it

Fausto Galvan
Digital Forensics Expert and Law Enforcement agent
Public Prosecutor office at the Court of Udine (IT)
galvanfausto14@gmail.com

Marco Alvise De Stefani
CEO of Synaptic Srls and Digital Forensics Expert
destefani@synaptic.it

Sebastiano Battiato
Full Professor of Computer Science
Department of Mathematics and Computer Science
University of Catania (IT)
battiato@dmi.unict.it

Abstract

IoT technologies pose serious challenges to digital Forensics. The acquisition of digital evidence is hindered by the number and extreme variety of IoT items, often lacking physical interfaces, connected in unprotected networks, feeding data to uncontrolled cloud services. In this paper we address "Information Quality" in IoT Forensics, taking into account different levels of complexity

---

*Although this contribution has to be considered as an outcome of an ongoing joint research, the single paragraphs should be mainly attributed as follows: 1-3 to Federico Costantini, 4 to Marco Alvise De Stefani , 5 and 6 to Fausto Galvan, 7 to Sebastiano Battiato.

and included human factors. After drawing a theoretical framework on data quality and information quality, we focus on forensic analysis challenges in IoT environments, providing a use case of evidence collection for investigative purposes. At the end, we propose a formal framework for assessing information quality of IoT devices for Forensics analysis.

Keywords: *Information Quality Assessment, Digital Forensics, IoT Forensics, Digital Investigations.*

1   Introduction: challenges in IoT Forensics

In the last twenty years, we have been witnessing the advancement of a forensic science known as Digital Forensics, whose aim is to develop a rigorous methodology for retrieval, collection, and analysis of digital evidence[1] One of the greatest challenges in this discipline is to keep up with the speed of technical innovation that, as we know, in the digital field is higher than in any other sector. In recent years, indeed, the analysis of such evidence has required demanding effort, especially due to the pervasive use of two technologies: cloud computing and artificial intelligence. On the one hand, the virtualization of resources hinders the validation of the source, the accuracy of the analysis, and the integrity of the results, since *"evidence can reside everywhere in the world in a virtualization environment"*[68]. On the other hand, the background of a decision taken by artificial intelligence systems lacks transparency, in that these systems' behaviour is as unpredictable as "black box"outcomes.

Currently, forensic analysis of such systems has been addressed in an effort to improve pre-existing methodologies. "Cloud Forensics" can be defined as *"the application of Computer Forensics principles and procedures in a cloud computing environment"* [59] whereas "explainable artificial intelligence" (XAI) aims to develop a suite of techniques that, bringing more transparency into the reasoning process, allows the validation of the reconstruction of external events [1]. Despite the promising results currently achieved in this way, such "disruptive technologies" require not only new technologies but also new approaches which have not been completely deployed yet.

We can observe the same evolution in the Internet of Things (IoT). The term was originally coined in 1999 with specific reference to RFID (Radio-Frequency

---

[1]Digital Forensics has been defined as *"the use of scientifically derived and proven methods toward the preservation, collection, validation, identification, analysis, interpretation, documentation, and presentation of digital evidence derived from digital sources for the purpose of facilitation or furthering the reconstruction of events found to be criminal, or helping to anticipate unauthorized actions shown to be disruptive to planned operations."* [58].

Identification) technologies [4] and soon overcame more general expressions such as "ubiquitous computing" [73], "pervasive computing" [34] and "ambient intelligence". Now it is commonly used to designate *"a global infrastructure for the information society, enabling advanced services by interconnecting (physical and virtual) things based on existing and evolving interoperable information and communication technologies"*, according to a definition given by ITU in 2012 [45, 30]. A previously only imagined scenario now characterizes everyday life. Today millions of devices — of many different types, models, and versions — are connected in an extensive infrastructure, exchanging enormous quantities of data, and it can be foreseen that this trend will increase dramatically in the future. According to a report by Ericsson, there could be 3.5 billion IoT connections in 2023, 2.2 billion of them in North-East Asia alone [17].

Forensic analysis of digital evidence in the IoT environment poses several challenges [35, 36, 80, 54]. Indeed, data are spread across an undetermined (e.g. in their type, number, and location) set of connected devices [79]; machines present different kinds of security vulnerabilities, so in being extensively exposed to hijackers, communications can be unprotected or even unencrypted, allowing third-party manipulation; and storage units often do not grant secure access to archives. Furthermore, due to the high interdependence among devices, any anomaly can spread rapidly in an IoT ecosystem and flood outwards; hence serious crimes, with great damage of exponential scale, could be committed without leaving any trace. As a matter of fact, methods tested as valid for isolating devices in "chain of custody"[2] situations, as in "classical" digital Forensics, are not effective in the IoT environment, due to the constant connections and deep interaction among devices.

IoT Forensics analysis requires both cutting-edge technological solutions and new methodological approaches in order to grant integrity, authentication, and non-repudiation of digital evidence[3] extracted from the acquired data. Three specific reasons can support such claims: objective, subjective, and transactive.

First, objective: the greater the quantity and diversity of evidentiary sources, the less trust can be put on their quality. With a huge amount and variety of sources to analyse, there is a great impact on the extraction of data. If many different kinds of devices are interconnected, it is likely that they must be approached with different standards and, conversely, it is unlikely that the results can be efficiently aggregated.

Second, subjective: the more data that are available, the harder it is to manage

---

[2]The set of procedures aiming to guarantee the integrity of evidence is called "chain of custody". It encompasses different phases and requires operations to be logged in order to ensure transparency and trust. See the "Convention on Cybercrime" signed in Budapest in 2001 (n. 185 CoE).

[3]A new and interesting perspective is brought by decentralized ledger technologies, as recently pointed out in an interesting paper by Brotsis [14].

them, even if deploying cutting-edge Forensics tools. Indeed, the quantity of evidence has a considerable influence on the computational effort required for information retrieval. Selection of relevant information and the processing of results can be very demanding, also considering that, especially in ongoing trials, the usefulness of the outcome strictly depends on its timeliness, and the analysis of an IoT ecosystem requires a consistent effort by Forensics consultants.

Third, transactive: the more advanced the technologies, the higher the expertise required for their analysis, and the lower the average understanding capacity by non-experts. Digital Forensics, indeed, as with any other forensic discipline, requires not only a solid background, rigorous methodology, professional experience, competences, skills, and up-to-date technological tools, but also the ability to support findings with convincing arguments. Evidence, even of a digital nature, needs to be not only acquired but also discussed with interlocutors who are not experts in the specific fields: judges, police officers, lawyers, and other consultants. If not plainly explained, the most grounded evidence might have no impact even if data were to be perfectly available, accessible, and genuine.

Digital Forensics analysis, and thus IoT Forensics, cannot be addressed without considering the three issues described above, which encompass the general problem of the influence of human factors in the transmission of knowledge. Of course, any forensic discipline is grounded in scientific methodology, and its assumptions are commonly justified by technological standards, established protocols, and widely accepted best practices. Moreover, the admission of evidence in court is ruled by adjective procedures, which confer strict obligations on all parties (judges, prosecutors, defendants, police officers, witnesses, consultants). Even so, it still is important to underline that the purpose of evidence is to be discussed and decided upon. In other words, both validity and efficacy of proof are noteworthy. The fact that evidence is digital, thus involving specific technologies and expertise, should not be allowed to undermine the issues concerning their impact in courts.

In this paper we explore IoT Forensics in terms of information quality (henceforth, IQ), offering a theoretical model which includes both data quality (henceforth, DQ), technological resources and human aspects. To that aim, we intend to proceed as follows: firstly, we provide an overview on DQ and IQ concepts, focusing on IoT issues; secondly, we focus on the concept of IQ we intend to explore in this contribution, including it in a wider theoretical framework; after that, we analyse the practical concerns, expressing the issues pointed out in a previous part; consequently, we propose a model for assessing IQ in the IoT environment and represent it in a formula. In conclusion, we express our final evaluations and draw paths for future research.

2 Data, Information, DQ and IQ: an Overview

"Data" and "information" are two deeply intertwined concepts whose core and mutual difference have been extensively debated among different approaches and disciplines. On one side, a pragmatic view can be epitomized by technological standards. In this sense, according to [39] "data" is defined in terms of "information", while "information" is considered a sort of "knowledge"[4]. Conversely, a remarkable example of a theoretical perspective can be found in "Philosophy of information", an approach developed in last the twenty years [25, 26], according to which "information" is described as a sort of "meaningful data" while "knowledge" is theorised as "accounted information" [28]. This latter proposal allows to shape a thorough vision which encompasses traditional ontology and contemporary epistemology and it is able to include also the social processes of transmission and sharing. In other terms, reality and knowledge are perceived and converted in "meaning" by agents — if humans, through interpretation, if machines, thanks to computation[5] — and then shared among each other regardless of their organic or artificial nature. In this view, we, as humans, are informational agents populating an ecosystem called "infosphere" as well as animals and computers [27].

In the legal field, the intricate relationship between reality and knowledge has always challenged the acquisition and admission of evidences. Indeed, in criminal proceedings every legal system entitles authorities with investigative powers for the collection of sources of evidence, which are submitted to judges according to specific procedures in order to bring decisions based upon them. Likewise, evidences are debated in civil trials where parties are equally enabled — at least in general — to produce facts or circumstances in support to their arguments or counter-arguments. Provided that an evidence — regardless the fact that it is incorporated in a digital support or not — can be qualified a sort of "information" , the problem of IQ becomes crucial for the application of law. In other words, it is important to establish transparent standards for the acquisition and admission of evidences in order to create an environment suitable for celebrating fair trials. This is essential in the case of digital forensics — especially in IoT forensics, as anticipated in the introduction — since the evidence is electronic, thus not engraved in a physical object.

---

[4]Indeed, "data" is defined as: "reinterpretable representation of information in a formalized manner suitable for communication, interpretation and processing", while "information" is defined as: "information processing knowledge concerning objects, such as facts, events, things, processes, or ideas, including concepts, that within a certain context have a particular meaning" [8].

[5]In this paper we cannot delve the difference between interpretation and computation. Our intention is to draw an abstract model suitable to describe the process made by the human mind and by an artificial agent.

2.1 Data and DQ

It has been said that data are the bridge that links the physical realm and the cyber world [48]. The life cycle of data has evolved due to recent technological innovations, as information is now processed both by humans and machines. Since IoT technologies allow an extensive, continuous, and direct flow of information among devices, the problem of DQ is crucial, especially if the interaction is not filtered by human supervision. Provided that the IoT is intrinsically untrustworthy due to the reasons mentioned above, this can lead to unpredictable consequences of an exponential scale, but the IoT also can generate anomalies which are also unperceivable to users, especially humans, thus hampering the possibility of arranging countermeasures or remedies.

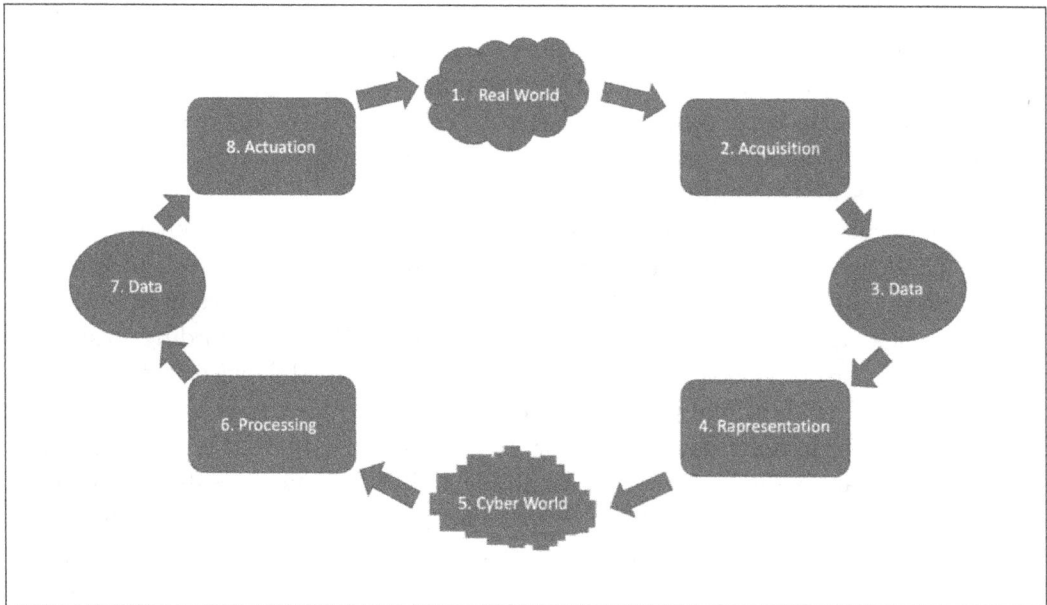

Figure 1: Representation of IoT scenario

As shown in Fig.1, the pipeline that allows one to collect a reliable set of information from a general IoT scenario starts with the acquisition of a set of data from the "real world". This could be done by a group of sensors, or by automated collecting procedures provided by the data owner, depending upon their data policy. This set of data is represented to the "cyber world" and then processed and re-introduced into the environment.

Although the concept of DQ has been crystalized by technical standards, being defined in ISO/IEC 25012:2008 [39] *"the degree to which the characteristics of data satisfy stated and implied needs when used under specified conditions"*,[6] there is wide debate over the scope and reference of the "fitness for the (intended) use of data" [21, 22, 7, 75].[7] Specifically, scholars have discussed the classification of such feature, among which some 159 components have been counted [48]. Table 1[8] shows a comparison of the different classifications.

In this paper, we make a methodological choice, preferring IQ over DQ in our approach, for three basic reasons: (1) the extreme complexity of IoT environments, as described above, in which data are just a part of a complex pipeline, along with devices and processes; (2) the possibility not only to include different kind of content — such metadata — but also to take into consideration the context of the devices, thus enabling a better representation of the complexity in the analysis; and (3) the possibility to include human interaction in the analysis of forensic evidence, since the meaning of data becomes as relevant as their source and the process performed to obtain it.

2.2 Information and IQ

Since "information" has a broader meaning than "data", as we argued above, the scope of IQ is wider than of DQ. As for the latter, in IQ scholars have proposed different criteria of classification — distribution, heterogeneity, and autonomy — which allow one to establish six different types of information systems (monolithic, distributed, data warehouses, cooperative, cloud, and peer to peer) [8]. Yet, one of the most interesting features of information is that it can be directly connected to the quality of the decisions that are based upon it. In this sense, an agent — either human or artificial — is influenced not only by shortage or by overload of information, but also by its quality. IQ, in short, is crucial for the outcome of the process, that is the utility of the decision in itself. Under this view, we can observe that some of the issues in IQ are the same as in DQ, for example in defining the dimensions under which quality can be addressed. In the field of Forensics analysis, two further profiles come into consideration if we consider the interaction between the devices examined and the Forensics expert. Indeed, the latter is entitled to bring

---

[6]As stated by Batini and Scannapieco in [8], quality in general has been defined as the "totality of characteristics of a product that bear on its ability to satisfy stated or implied needs" ISO 9000:2015 [38]. See also the definition of "data quality" in ISO/TS 8000-1:2011 [44].

[7]Also called "fitness for (intended) use" [47]; "conformance to requirements" [20], or "user satisfaction" [71].

[8]The first four columns of the table represent different classifications proposed by four contributions mentioned in [8] which did not include [67], probably because it was published afterwards. Other models mentioned by Batini and Scannapieco are not tackled in this contribution being not relevant for IoT forensics [46, 12, 50, 56, 61].

(Batini and Scannapieco 2016)				
(Batini, Palmonari, and Viscusi 2012)	Theoretical approach (Wand and Wang 1996)	Empirical approach (Wang and Strong 1996)		(Shamala et al. 2017)
Accuracy	Accuracy	Intrinsic	Accuracy	Accuracy
-	-		Believability	Believability
-	-		Objectivity	-
-	-		Reputation	-
Completeness	Completeness	Contextual	Completeness	Completeness
-	-		Relevancy	Relevancy
-	Timeliness		Timeliness	Timeliness
-	-		Value added	-
Redundancy	-		-	-
-	-		Appropriate amount of data	Amount of data
-	-	Representational	Interpretability	-
-	-		Ease of understanding	-
-	-		-	Objective
-	-		Concise representation	Concise representation
Consistency	Consistency		Representational consistency	Consistent representation
Usefulness	-		-	-
Trust	-		-	-
-	Reliability		-	Reliability
Accessibility	-	Accessibility	-	Availability
-	-		Accessibility	-
-	-		Access security	-
Readability	-		-	-
-	-		-	Understandability
-	-		-	Verifiability

Table 1: Comparison of different classifications of features in DQ.

the Forensics analysis to court and thus embodies the human decision maker for the filtering of data. Specifically, we have to consider (1) how information is collected by the decision maker and (2) how information affects the decision. Concerning the first aspect, it is noteworthy that IQ represents a kind of "meta-information", which should be taken into consideration by the agent in the decision. Here two aspects need to be measured: complacency[9] and consensus.[10] With regard to the second aspect, not only does the interpretation of information come into play (the accuracy and the depth of understanding of the agent) but also pre-existing factors — such as the agent's beliefs — or contextual aspects such as cognitive overload or time constraints.[11]

3 Theoretical background of IQ in IoT Forensics

According to "Philosophy of Information", as outlined above, "information" has three ontological statuses, which take into account the seminal studies of Weaver[12]: "information *as* reality", for example the electrical signal, which is transmitted regardless of the message contained; "information *about* reality", such as information about natural phenomena, which can be true or false (hence in philosophical terms can be said to be "alethic") [51, 23]; and "information *for* reality", which conveys instructions or algorithms to one or many recipients. Each of them has to be considered separately, since they refer to different aspects of reality.

Therefore, IQ can be studied under three different views.

- **Quality in "information *as* reality"** is the most common feature and emerges for example in the traditional problem of reducing noise, distortion, or losses in signal transmission. In this sense, it measures the affordability of the means implemented to transfer information;

- **Quality in "information *about* reality"** is concerned with the dissimilarity of information to the facts to which it refers. This concept is related to the meaning of information, or semantics [10, 6]. Thus, quality in this case

---

[9]This feature measures *"the degree to which information on IQ is ignored"* [8].

[10]This feature measures *"the level of agreement within a group with respect to a preferred choice"*[8].

[11]These two aspects correspond respectively to the distinction between *"the purpose/s for which some information is originally produced (P-purpose) and the (potentially unlimited) purpose/s for which the same information may be consumed (C-purpose)"* [25].

[12]In the original exposition of the theory of communication, such concepts were expressed as different "levels", respectively as "technical", "semantic", and "influential" [72]. Instead, cybernetics defined "technological", "natural", and "cultural" information [11].

measures the reliability of the information provided in representing the related events;

- **Quality in "information *for* reality"** has to be addressed when processes present inconsistencies, loopholes, or conflicts. This concept involves further processes of information, for example when it is shared with others. Hence, in this respect it measures the trustworthiness of the agent who receives information or, generally, of those involved in further processes.

In Digital Forensics, the problem of IQ has received special consideration under different perspectives. Here we can deploy the same tripartite analytical model to draw a comprehensive framework.

- The first kind of quality is relevant in order to preserve the integrity of the collected information, since experts developed the concept of chain of custody and achieved a broad consensus on best practices which have been codified in technological standards [40, 41, 42, 43];

- The second type of quality is concerned about the trustworthiness of the representation of events, which has to be verified with other sources of evidence [29];

- The third sort of quality is involved in the discussion of evidence among parties (inquiring authorities, defendants, judges, Forensics experts). As we know, trials have to proceed according to precise rules which establish specific requirements for admissibility and the burden of proof. Here also, external variables can make a difference, such as personal competences of the agents involved, "soft skills" (argumentation abilities, trial strategies), cost of analytical tools, and available time.

From a theoretical perspective, since devices in the IoT are constantly sharing information with each other, IQ is a highly complex issue. Indeed, it has to be considered not only as a property of a single device but as a feature of the whole ecosystem in which every item is immersed. Specific concerns emerge in each considered aspect:

- **Quality of "information *as* reality"**. It is difficult to isolate a single device or crystalize a specific piece of information. The boundaries of relevance are blurred. This aspect is, not without reason, widely discussed by experts[74, 18];

- **Quality of "information *about* reality"**. It is problematic to detect a specific source, to trace the chain of interactions, or to measure the influence of a single item in shaping the representation of an event. It is commonly true that correlation is not causation; however, the IoT is, above all, a matter of correlation. Here in the IoT is where quality of information really faces uncertainty[52];

- **Quality of "information *for* reality"**. This is the most difficult aspect of IoT Forensics. Under this perspective, the human factor plays a part along with technological variables, as shown in digital Forensics.

The classification outlined above is represented in Table 2.

Ontological statuses of information	Quality of information; level of analysis	Quality of information in digital Forensics	Quality of information in IoT Forensics
Information *as* reality	Traditional theory of communication	Chain of custody	Relevance
Information *about* reality	Consistency with other represented facts	External validation with other sources of evidence	Uncertainty
Information *for* reality	Logical coherence	Adjective rules (admission and burden of proof)	Accountability

Table 2: IQ tripartite analysis and IoT issues.

Accountability is noteworthy in social processes, hence its legal relevance. This concept finds its roots in management theory, where one can find this general definition: *"to be accountable for one's activities is to explicate the reasons for them and to supply the normative grounds whereby they may be justified"* [33].[13] Accountability is crucial in fiduciary positions held on behalf of third parties, which are not directly involved in decisions that an agent has to make. The third party has the power to set a certain policy under which decisions have to be made by the agent, who is required not only to act according to said policy but to explain the reasons for her/his choices [49].

4 Practical issues in IoT Forensics

Accountability is fundamental in Digital Forensics, especially when solutions cannot be granted with certainty. When opinions are debatable, it is very important to be transparent about the methods adopted and to conveniently share the results obtained.

---

[13]In corporate management studies, accountability is crucial in decisions by corporate boards [64].

For example, during the bitstream copy of a hard disk it is very important to be transparent about the handling of the hard disk, the write-blocker and imaging software used, the log of the imaging process, the hash of the image, and so on; with this transparency, we know how the process is performed and thus can assess the IQ of the results obtained.

In IoT Forensics the observer faces three main problems related to accountability. She or he or needs to:

1. Acquire only data that are related to the case and that can be of interest. Otherwise, in IoT it is very easy to flood a case with billions of data points that after analysis come out to be useless. Selection has to be justified;

2. Assess the degree of uncertainty of the information, which must be a genuine representation of the reality. Such evaluation has to be clarified in its tenets;

3. Choose the tools of acquisition. Such a choice has to be explained, moreover, if for technical reasons the acquisition cannot be repeated in the future under the same conditions.

The main feature of an IoT network is the possibility, while performing a given tasks, and when necessary, to take advantage of information exchange with other neighbouring elements, regardless of the fact that they could belong to a different IoT ecosystem, possibly set up for a different purpose. Over the years it has become clear that, while it was originally conceived as the beginning of new opportunities for increasing the efficiency of services — thus to benefit private users and industries — this technology now generates a "virtual environment" containing a huge amount of information which, if needed, can be used as a source of evidence in a forensic scenario. In the following part, we observe that the main task of IoT Forensics, namely the collection of evidence, is challenging for different reasons.

First, the extreme variety of IoT items, with proprietary or undocumented protocols, often without physical interfaces, hinders the direct extraction of evidence from devices.

Second, IoT devices ceaselessly send information to "their own" cloud service providers. Thus, obtaining evidence from a cloud service involves other challenges, both technical (cloud Forensics) and legal (interacting with foreign companies in different legal frameworks).

Third, evidence could be manipulated by cloud service providers. And raw data fed by IoT devices are analysed, parsed, and stored in databases and servers, adding yet more layers of interpretations and classifications. Those processes can aggregate data but also deteriorate them, thus weakening the quality of the acquired information.

To explain such issues, we provide an example. Let's assume that we need to acquire the complete geolocalization history of a specific account, from a cloud service provider (i.e. Google Timeline History[14]).

Let's suppose also that we have the account's credentials (i.e. username and password): we can use forensic software (i.e. Oxygen Forensic Cloud Extractor[15]) to acquire the geolocalization history.[16] After the acquisition, we can explore and analyse the collected data: we will find a very clear and detailed set of information, spanning years and neatly organized. The typical forensic software user interface for this type of data consists of a world map we can freely zoom and search, a time filter, and coloured pins and lines that represent geolocalization information and probable movements.

This variety of sources can lead us to assume that the quality of data is very high, but after an additional analysis we may find that we don't know where the information comes from. We know that the service provider could acquire geolocalization information from a great range of sources related to that specific account: a smartphone's GPS; EXIF data of images; a cellular network radio tower; WiFi known geolocalization; near-field communication; fitness apps and arm bands; Uber rides; IP geolocalization; smartwatches; navigation apps; reviews of shops or restaurants; public transportation tickets; and many other IoT sources. The timeline can even be manipulated by the user through the cloud service web interface (i.e. in Google Timeline we can easily add a new geolocalization or change or delete an old record).

If we focus on a specific day, we only have the alleged geolocalization history: some points (consisting of longitude, latitude, and time stamp) and lines that connect those points to speculate movements, but we are not able to discover where each point comes from. Even if in some metatag there is information about the source of the info, we can access only the parsed data; we don't know if the source is reliable or if the data has been manipulated, compromised, or not properly interpreted. The original data may not even be accessible (i.e. if the geolocalization comes from a fitband's GPS, probably we can't physically connect to the device; we can acquire the paired smartphone and analyse the fitband's app's databases, but even that isn't the original data).

In this specific example (Google) we can also acquire the same data through the Google Takeout procedure in two different formats: JSON and KML.

We can try to understand what happened in reality only by comparing these

---

[14]https://www.google.com/maps/timeline.

[15]https://www.oxygen-forensic.com/en/oxygen-forensic-cloud-investigator-ofci.

[16]We presume that the forensic software is compliant with all digital Forensics procedures and standards.

three sources (Timeline and JSON/KML Takeout) and interpret the results with the experience and the reverse engineering in similar cases.[17]

5 Proposal for a theoretical model of Information Quality Assessment in IoT Forensics

The evaluation of IQ in a set of heterogeneous data is not a trivial task, as the previous sections have shown. The various steps of the assessment always hang on different factors, and many of these cannot always be precisely quantified but only be estimated[18]. In addition to these basic difficulties, in common for every environment, as said before the IQ needed in a Forensics scenario must undergo a further filter, represented by a "check for admissibility", which allows the acquired data to be part of a trial procedure.

5.1 From the theoretical framework to a mathematical approach

A possible approach for such an issue could be based on the separate verification of IQ in two different fields, as shown in Figure 2. The information must be proved to be at the same time admissible as evidence in a trial — thus properly acquired and stored according to the latest and best practices in the field [40] — and reliable as a collection of data (with a low amount of noise, readable, understandable, saved in a known format, etc.); thus it will serve as input for the Forensics analysis. This method, although interesting, is affected by two main risks: wasting time with one of the two controls when the outcome of the other is negative, and rejecting key evidence which, for some reason, is not proven to be reliable as high quality IoT data (e.g. a noisy image which, considered together with other evidence, could have an important meaning).

This last observation leads to the development of another approach, where the two aspects of the extracted information (since they are, at the same time, "rough" data and "evidence") are spread all over the items of a sum of proper features.

Following the above considerations, we implemented the theoretical framework drawn in the previous section concerning the classification of the requirements of IQ

---

[17]In other words, the same sporadic geolocalization (down to the cm) on different days, with a large uncertainty range, probably means that the smartphone was connected to a cellular radio tower.

[18]We cannot forget the noise coming with the data flow, which must be carefully identified and removed. Of course, such a process has to be performed very cautiously since it may cause the definitive loss of precious data.

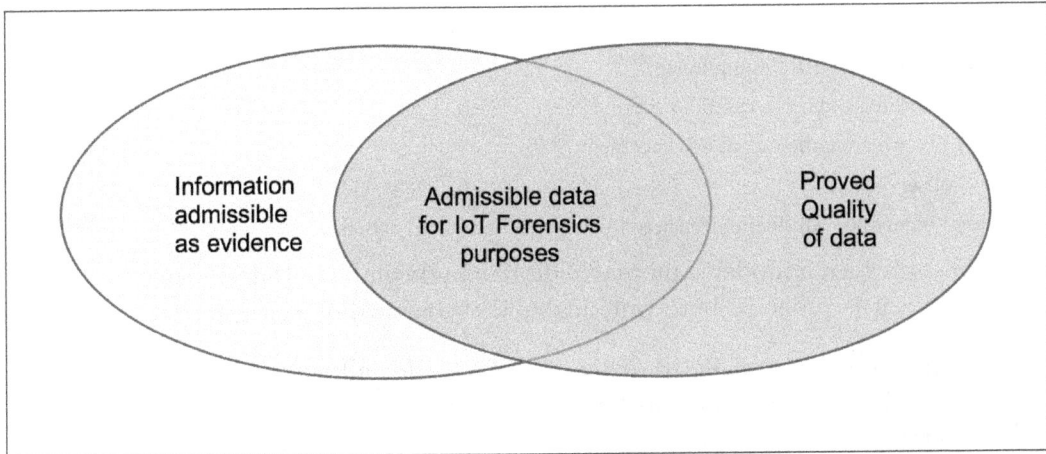

Figure 2: Requirements for IQ assessment. To be admissible for forensics purposes, the acquired information has to be, at the same time, readable and understandable as a digital data, and acquired and stored with respect to the best practices in the specific forensic field.

in an IoT environment. For sake of clarity, we consider only the categories of IQ features shown in [69].

In our fictional investigative scenario, all the collected digital evidences come from a set of n IoT devices. Our model takes into consideration several factors for establishing the IQ of the information extracted from every source, introducing a percentage coefficient that we named IQA (information quality assessment), as follows:

$$IQA = \frac{\sum_{i=1}^{n}(DTC_i + DST_i + CS_i + CM_i + SR_i + PC_i + TDA_i + OT_i + OS_i)}{9n} \times 100 \quad (1)$$

where:

- i = i-th device;

- DTC = device technical status;[19]

- DST = device security status (confidentiality, integrity, availability, ...);[20]

- CS = cloud service security status;[21]

- CM = cloud service manipulation of raw data;[22]

---

[19]The fact that the device is working properly (both hardware and software) and the currency of the software.

[20]The level of protection deployed in the device, both in hardware and software.

[21]The level of cyber protection of the cloud where data are stored, even temporarily, and of the transmission of data to and from the cloud.

[22]The possibility that in the cloud the data are processed and changed somehow.

- SR = source reliability;[23]
- PC = privacy (GDPR) compliance;[24]
- TDA = technical data accessibility;[25]
- OT = observer technological advancement;[26]
- OS = observer skills;[27]
- allowed values are all decimal between 0 = "bad" and 1 = "good";

Although such a model is more exploitable and complete than the one described in Figure 2, it is prone to both practical and theoretical objections:

1. It may not be possible to define each factor for all types of IoT devices. For example, the one named as CS seems to imply that the device is always connected to its environment, but some devices can also (or exclusively) be used offline;

2. The data needed to define these variables may not be accessible, not only technically, but because of the restrictions imposed by the companies that own them. For example, let's suppose that an investigation in Italy requires data related to a journey of a car produced by a foreign company. Let's think that, for some reason, the only copy of these data are stored in a server located the country where the company is based: it could be very complicated to reach such a source and collect data from it;

3. The terms of (1) belong to heterogeneous aspects, some technical, others mostly connected to the human factor, or depending on abstract concepts or that can be changed by the legislator. Therefore, it is arbitrary to estimate their mutual interactions, thus their contribution to (1). Questions like the following could arise: is it right that an increase of 0.1 in the device technical status has the same influence on (1) as the increase of 0.1 in the privacy (GDPR) compliance? What does it mean to increase, i.e. 0.15, in one of the items? Is the idea of "good" OS the same in Italy and the other countries?

---

[23]The possibility that the observer possesses additional information about the source which originates the data, i.e. the IQ level obtained in a previous investigation of data originating from the same source.

[24]The compliance with data protection regulations, for example in terms of data protection by default or by design. This requirement is related to the GDPR in the EU, yet of course it depends on the fact that in the legal system taken into consideration there is specific legislation.

[25]The availability of technical specifications regarding the device and the format.

[26]The analytical tools used by the observer to collect and process evidence.

[27]The degree of experience, abilities, and skills achieved by the observer before the observation.

4. In some cases it could be useful to set up an IN/OUT threshold value for IQA, a value under which any data can be discarded. This is an investigative choice that should be carefully considered and clearly explained.

All the aspects of the acquired data deserve to be explored in depth, without being mixed, and maybe confused or underestimated. In addition, thus giving an answer to objection number 4), we can't forget that in a forensic scenario all information must be taken under consideration, regardless of its quality, so dividing data depending on their belonging to one of the three categories assures us that we will not waste any details.

We suggest that the issue raised in point 3) can be tackled with the following precautions:

a) Pondering each element to add in the formula, depending on the contribution of the item to the whole evaluation;

b) Separating the items into different areas. Consequently, for every group of IoT evidence we would have different tags, yet scores can remain separated.

Concerning this last observation, we can divide the factors of the formula accordingly with two different kinds of taxonomies, whose adoption depends on the utility that it is more suitable in the trial. Of course, such a choice is not mutually exclusive, since in the trial every aspect can be used in argumentation. Indeed, we can highlight the kinds of information ("as", "about", or "for" reality) or the layers involved, as explained in the following paragraphs.

The two different kinds of taxonomies allow us to fine-tuning the model provided above into two opposite directions. In the first one we can aggregate the IQA factors into the four categories shown in Table 3, thus being able to split (1) in three formulas, each of which takes under consideration a different category of the "Philosophy of Information":

$$IQA_I = \frac{\sum_{i=1}^{n}(DTC_i + DST_i + CS_i)}{3n} \times 100 \tag{2}$$

$$IQA_{II} = \frac{\sum_{i=1}^{n}(CM_i + SR_i)}{2n} \times 100 \tag{3}$$

$$IQA_{III} = \frac{\sum_{i=1}^{n}(PC_i + TDA_i + OT_i + OS_i)}{4n} \times 100 \tag{4}$$

where:

- IQA_I = information *as* reality
- IQA_{II} = information *about* reality
- IQA_{III} = information *for* reality

Categories	Philosophy of information	Requirements
Intrinsic	Information *as* reality (relevance)	DTC
Contextual		DST
		CS
Representational	Information *about* reality (uncertainty)	CM
		SR
Accessibility	Information *for* reality (accountability)	PC
		TDA
		OT
		OS

Table 3: Synopsis of IQ requirements and information categories.

In the latter point of view, (1) can be shaped considering that an IoT environment is characterized by the presence of different layers: physical perception layer, which includes sensing or moving; network layer, encompassing processing and transmission; and application layer, which corresponds to the services provided [48].[28]

Each factor can be considered under three aspects, depending on the layer taken into consideration. As an example, DST — the technological security of the single device — can be evaluated under the physical aspect (i.e. protection case), the network security (i.e. encrypted transmission), and the application layer (i.e. password-protected interface).

---

[28]Such layers are defined as follows[76]: "*A physical perception layer that perceives physical environments and human social life, a network layer that transforms and processes perceived environment data and an application layer that offers context-aware intelligent services in a pervasive manner*".

Categories	Requirements / layers		
	Physical perception layer	Network layer	Application layer
Intrinsic	DTCp	DTCn	DTCa
Contextual	DSTp	DSTn	DSTa
	CSp	CSn	CSa
Representational	CMp	CMn	CMa
	SRp	SRn	SRa
Accessibility	PCp	PCn	PCa
	TDAp	TDAn	TDAa
	OTp	OTn	OTa
	OSp	OSn	OSa

Table 4: Synopsis of IQ and IoT layers.

Therefore, it is possible to draw the synopsis of Table 4,[29] that drives us towards the definition of three other formulas that allow us to take under consideration this different approach:

$$IQA_p = \sum_{i=1}^{h} \left(\frac{\sum_{i=1}^{m_{p_i}} f_{p_j}}{m_{p_i}} \right) \tag{5}$$

$$IQA_n = \sum_{i=1}^{h} \left(\frac{\sum_{i=1}^{m_{n_i}} f_{n_j}}{m_{n_i}} \right) \tag{6}$$

$$IQA_a = \sum_{i=1}^{h} \left(\frac{\sum_{i=1}^{m_{a_i}} f_{a_j}}{m_{a_i}} \right) \tag{7}$$

where:

---

[29]Of course, OT and OS are related to the observer, which is a human being. This means that the factors represent the technical apparatus deployed by the consultant and the skills acquired in each field. For example, a consultant could have a very advanced tool for device analysis (OTp), but not the same ability to guarantee the security of the connection (OTn) or the skills to evaluate how services are provided (OTa).

- h represents the number of devices taken into consideration in the investigation;

- i = i-th device;

- m_{p_i}, m_{n_i} and m_{a_i} represent the number of factors that can be evaluated, respectively, in the physical, network, and application layer, for the i-th device. This is because, as clarified above, not all factors of Table 4 have to be present in the considered formula.

- f_{p_j}, f_{n_j} and f_{a_j} represent the single factor of Table 4 that can be evaluated (i.e. DTCp).

- IQA_p = information of physical perception layer;

- IQA_n = information quality of network layer;

- IQA_a = information quality of application layer.

The considerations set out in this Section, make us confident of being able to make the "leap" towards operational realities, and to address the difficulties that surely could emerge in an investigation. In the next Section, we will expose two different case studies, that allow to appreciate how the terms abstractly defined in (1), and reused in (2), (3) and (4), could be practically evaluated and exploited.

6 Use cases of IQ in the IoT environment

As an application of all the above formulas in a real scenario, in the following section we propose two operational situation. Although both of them regard the issue of Information Quality Assessment, the first example is focused on the interpretation of the abstract terms in (1) in case of only one kind of digital evidence, still images and videos[30], whereas the latter, also proposed in [31], shows the powerful of the graphic visualization of the propose IQA assessment for an heterogeneous group of devices.

6.1 Case Study — 1: IQA assessment of images & video.

On a crime scene, a video surveillance system were seized by the law enforcement agents. Subsequent investigations proved that the recorded footages wasn't saved on site, but instead uploaded in real time on a server located in a foreign country. The rogatory procedures for the acquisition of the data allowed law enforcement agents to retrieve the requested source of evidence. Investigators need to assess IQA of the obtained digital evidences applying (1)–(4).

Before facing the exposed scenario, we move for a while from the meaning of IQ discussed in this paper, and we briefly try to focus upon the problem of determining the quality of a digital image, first by considering it "only" as a stand-alone digital representation of the real world, and then as a source of evidence that has to be

---

[30]One of the most useful sources of evidence in trials consists of visual documents [3].

acquired for a trial. To grasp the concept of digital image quality is not easy, due to the fact that it is not only a translation of the real world with "zeroes" and "ones" but also, or maybe mainly, a way to transmit emotions, a moment in life, and, of course, information. In the scientific literature this notion has various interpretations [8]:

- The subjective impression of **how well image content is rendered** or reproduced [63];

- The integrated **set of perceptions** of the overall degree of excellence of an image [24];

- An impression of its merits or excellence as **perceived by an observer** neither associated with the act of photographing nor closely involved with the subject matter depicted [66];

- The perceptually weighted combination of all visually significant attributes of an image when considered in its marketplace or application ('Camera Phone Image Quality (CPIQ)-Phase 1-Fundamentals and Review of Considered Test Methods (v. 1.10)' 2007).

Other definitions are given from different projects and points of view.[31] In addition, we also wish to propose our first version of a definition of image quality, which is the following:

Definition 1: *The quality of a digital image is represented by its "closeness", both numerically and semantically, to the real scene it was intended to represent, considering one (or more) proper metric(s).*

The background of Definition 1 is that the digital image formation pipeline, described in detail in the literature [53, 32], possesses all the steps shown in Fig. 1, translated for the specific field as in Fig. 3. For this reason, we intend the digital image quality as being inversely proportional to the "difference" (defining a proper measure of distance in this interdisciplinary field is absolutely not a trivial task) between the real scene and its digital version. In other words, asking if an image has a "good" quality means asking "how well" the image allows to reconstruct all aspects of the shot scene, both objective and subjective. Although complete and exhaustive for an image not to be used as an evidence in a trial, the kind of quality requested in Definition 1 is not enough robust for a forensic scenario, as we are going to clarify in the following.

Image/Video Forensics is a part of Multimedia Forensics [9], which is a forensic science that aims to validate the authenticity of images and videos by recovering

---

[31] Among others, see [2] and [78].

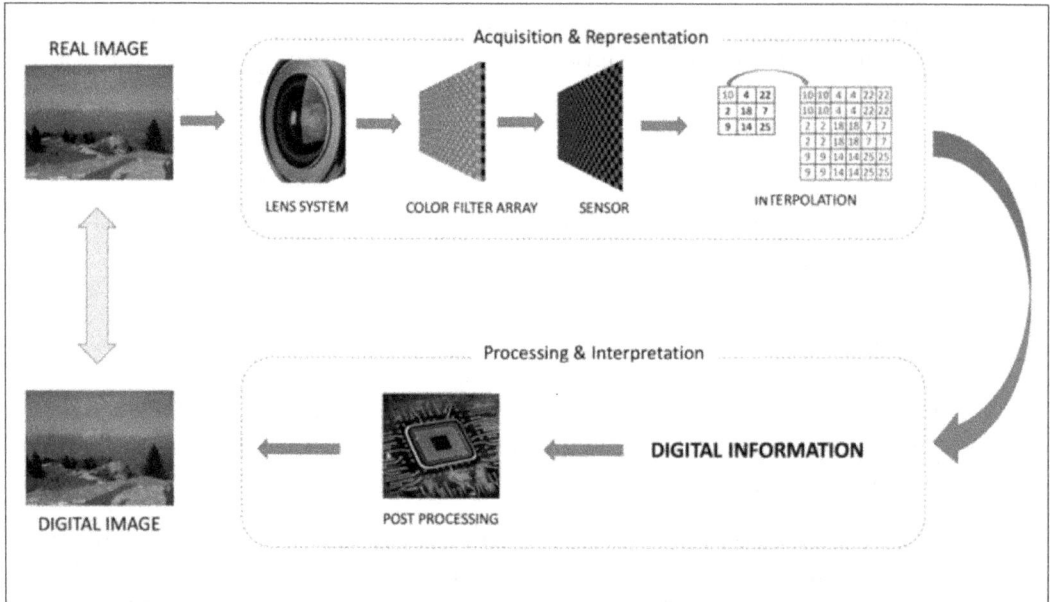

Figure 3: Steps of the formation of an image with reference to Figure 1. The real scene and its digital representation in an image are as close as the error (measured with a proper metric) between them is small.

information about their history [60] for reasons connected to an investigation or a trial, where they appears as evidences. This implies that in the pipeline devoted to define the quality of an image or a video in a forensic context, we must also consider if this file has been handled with respect to the "chain of custody" described by the best practices in the field [40]. This means paying attention to:

- Where the image comes from;

- How the image has been acquired;

- How it has been manipulated during the analysis;

- How is has been preserved before and after the analysis.

We can also argue that definition (1) raises another issue, since it is not operational. Indeed, it is not suitable to denote the complexity of forensics procedures as described above, being unable to include a multidimensional representation of the device, its exchanges of information with its environment and the extraction of data made by the observer. A step forward is needed.

Therefore, depending upon this different point of view, the definition of digital image quality in a forensic context changes in:

Definition 2: *The quality of a digital image considered as evidence in a forensic process is represented by its "closeness", both numerically and semantically, to the real scene it was intended to represent, considering proper metrics, and the recommendation exposed in the best practices about the identification, collection, acquisition, and preservation of digital evidence.*

After these considerations, recalling the operational question raised above and with reference to all the terms of (1), answering to the query exposed in the begin of this section having in mind definition (2) means being able to complete the following checklist:

Items	Meaning in the case of study
DST	Is the hardware set up, wired, and maintained taking into account the security rules corresponding to the best practices? Is the software devoted to managing the system provided with proper and up-to-date antivirus protection? Are the accesses to the system properly logged?
CS	Is easily obtained the information about the channel towards data are broadcasted? Is the entity from which data originate certified and reliable?
CM	Are data stored, even for a short period of time, in a repository that could be totally or partially accessible by some agent?
SR	Does the observer possess, or able to obtain, additional information about the source where the data originate?
PC	Are the footage and related metadata saved and preserved following the most recent GDPR precepts?
DTC	What is the technical status of the hardware part of the system? Is the software part of the system updated to the latest release?
OT	Is the observer recently acknowledged in some way as a valuable operator in the field?
TDA	Are the technical specifications regarding the device and the format of the obtained data easily available, or provided directly by the owner of the data?
OS	What is the expertise of the observer as a digital Forensics investigator?

Table 5: Checklist of issues emerging from IoT Forensics.

Facing an operative scenario with the exposed mathematical model, being able to assess numerical values to every term of (1) is crucial. Every type of evidence requires different approaches. In this table the checklist needed to calculate the IQAs from (1)-(4).

It is remarkable that definition (2) of image quality embraces a more pragmatic perspective. In adopting it, forensics experts need to be aware that this needs to fulfill a methodological further pledge of transparency — which can be reconnected with an obligation of "accountability", as stated above — thus requiring the development of an argumentative strategy aimed at avoiding misinterpretations.

6.2 Case Study — 2: IQA assessment for a set of digital evidences.

On a crime scene, a set of IoT digital devices were seized. Specifically, a smartphone; 2) the SIMCard inside of it; 3) a drone; 4) a smartwatch; 5) a laptop pc; 6) a smartTV. A Digital Forensics expert, before analyzing in depth these sources of evidence, must ascertain the IQ of every single device both individually and globally.

Comparing IQs of the devices and evaluating the overall IQ taking under consideration all the terms that compose (1) could be a difficult task. Indeed, the device manufacturers may not (or at least not yet) have made public the requested technical information or, even if released, these data could be not as detailed as necessary. In addition to that, information inside each device could be organized and stored in different ways, depending on the policies of the respective manufacturer. The level of these evaluations should be similar to what exposed in [16], [13] ad [57], where the file system, the shape and the format of the log files and other useful forensic clues are exposed in case of a drone, a smartTV and a smartwatch. After this kind of deep analysis, we could fill a table as Table 7 below[32], and then implementing (1), (2), (3) and (4), and thus generating a set of charts that allows to better insight the IQ of the examined sources of evidences. For the test we considered the following devices:

1. Smartphone Huawei model ALE-L21 (P8 Light), with Android 6.0, 2 Gb RAM, CPU Octa-core 1.2 GHz, kernek version 3.10.86-g33ff982;

2. Nano SIMCard 4G Telecom Italia year 2017;

3. The same model proposed in in [54];

4. The same model proposed in in [13];

5. IBM Thinkpad Edge E30, o.s. Windows 10, 8 Gb RAM, Intel i5 processor;

6. The same model proposed in in [57].

---

[32]The exact numerical value inserted in this example for every entry of Table 7, is not so crucial for the purpose of this paper, since were calculated after an evaluation made by the authors, and they could be subjected to significant variation depending upon a lot of reasons, e.g. changes in the laws, technical improvements, enhanced technical ability, and so on. Their purpose is allowing the calculation of the requested IQs and the graphical representation of the results.

device 1 smartphone		device 2 SIMCard		device 3 drone		device 4 smartTV		device 5 pc laptop		device 6 smartwatch	
DTC	0,56	DTC	0,93	DTC	0,97	DTC	0,91	DTC	0,39	DTC	0,89
DST	0,62	DST	0,12	DST	0,48	DST	0,16	DST	0,30	DST	0,82
CM	0,34	CM	0,17	CM	0,76	CM	0,60	CM	0,58	CM	0,85
SR	0,48	SR	0,76	SR	0,50	SR	0,98	SR	0,56	SR	0,18
PC	0,77	PC	0,82	PC	0,77	PC	0,19	PC	0,00	PC	0,98
TDA	0,55	TDA	0,60	TDA	0,21	TDA	0,44	TDA	0,26	TDA	0,65
OT	0,84	OT	0,07	OT	0,89	OT	0,80	OT	0,07	OT	0,89
OS	0,26	OS	0,88	OS	0,45	OS	0,64	OS	0,79	OS	0,31

Table 6: Numerical values of terms in (1), as evaluated by the authors for devices 1–6.

By applying (1), (2), (3) and (4) to all devices, with the data exposed in Tab.6 as input, we obtain the results listed in Tab.7: the IQA of the set of all seizured devices is about 62%, the device nr.4 is the one achieving the best result in terms of Information Quality, whereas device nr.5 bears the worst performance:

$IQA_I = 61,96\%$ $\quad$ $IQA_{III} = 54,74\%$ $\quad$ $IQA_{device2} = 59,49\%$ $\quad$ $IQA_{device4} = 89,79\%$ $\quad$ $IQA_{device6} = 66,68\%$

$IQA_{II} = 56,30\%$ $\quad$ $IQA_{device1} = 54,37\%$ $\quad$ $IQA_{device3} = 63,73\%$ $\quad$ $IQA_{device5} = 38,19\%$ $\quad$ $IQA_{tot} = 62,04\%$

Table 7: Results of the application of (1), (2), (3) and (4) to the seizured devices, using the data exposed in Tab.6.

The IQ can be clearly visualized also by a set of *radar* chart, which offers an immediate representation. In Figure 4, a set of evaluations is showed, considering both the total of the acquired staff and the single device. Subfigure a) represents a model of the best result that can be achieved: all the elements that compose the evaluation are at the maximum level, so the polygon is completely surrounded by the blue line. Subfigure b) shows at the same time the IQA of all the examined devices, and allows appréciating immediately the best result of devices nr.4 already highlighted. Subfigure c) shows together the IQA calculate with (2), (3) and (4), whereas in subfigures from d) to i) the performances of every single device is represented. Also from the comparison between these latter group of images, the peaking values of device nr.4 clearly appears among the others.

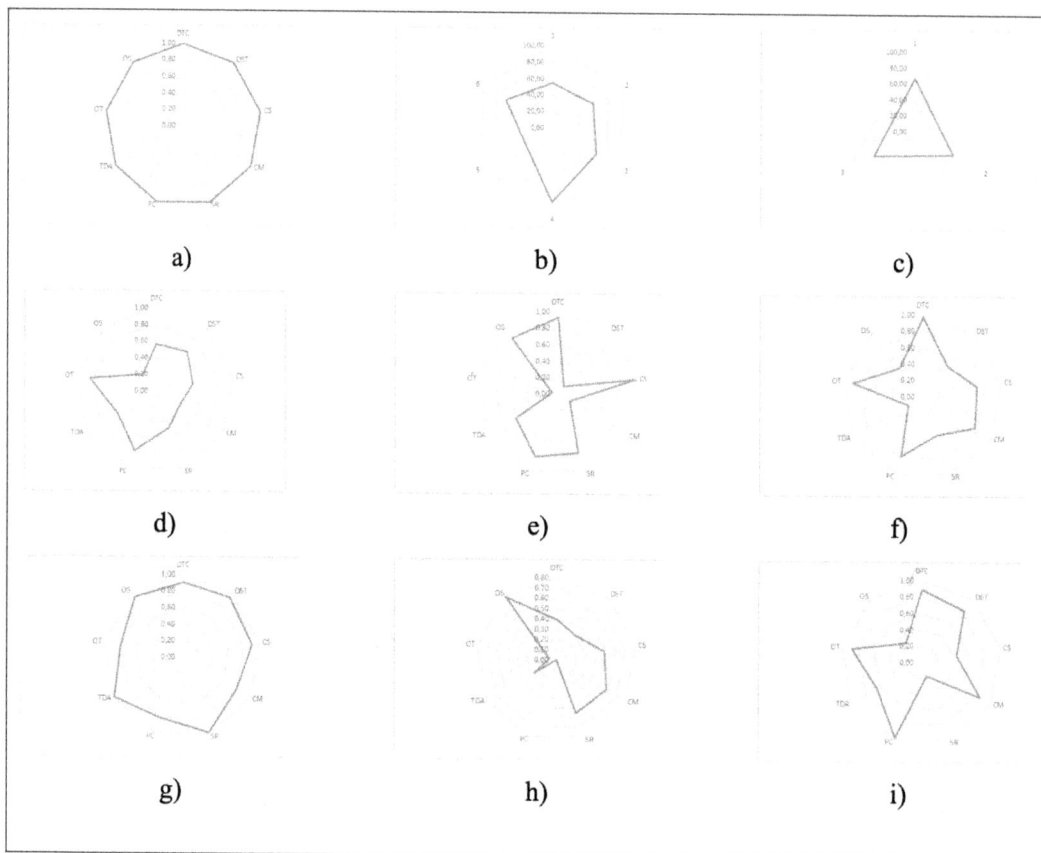

Figure 4: Graphic visualization of the outcomes of Case Study 2. The results of the IQA calculated by (1), (2), (3) and (4) become more intelligible with the help of this kind of charts, where the bigger the part of the inner figure is surrounded, the best is the achievement. In the above representation, a) was given as a model, and represents an example of the best result that a certain evaluation could achieve, since all the evaluated terms are at the higher level; b) is the IQA of all the devices showed together, that allows to highlight how, in the considered case work, the device 4 is the one with the best performance; c) shows at the same time the IQA_I, IQA_{II} and IQA_{III}; d) - i) are the charts referred to every term of the formulas of every devices, respectively 1 - 6. Also in this comparison confirms that, devices nr.4 obtains the best result.

7 Conclusions

Civil and criminal proceedings can be side-tracked from reaching their fair conclusion by mistakes or misinterpretations made by several actors: judges, defendants, prosecutors, police officers, consultants, witnesses. In this, digital evidence has become more and more important not only because it is increasingly widespread but also because of the ever-increasing skills required to manage it. It is crucial to improve the efficiency and the efficacy of evidentiary processes and to raise awareness of the peculiar nature of digital evidence for all the professionals involved.

Of the many challenges currently faced by digital Forensics, IoT technologies are of most concern, not only for practical but also for theoretical reasons.

From a theoretical perspective, many questions need to be deepened. Aside from the discussion of DQ and IQ approaches, it is remarkable that IQ is shaped differently depending on the level of complexity addressed (human to human, human to machine, machine to machine). Additionally, the problem of quality should be considered a fundamental issue in the design of IoT devices and ecosystems. Ethical concerns in this field should focus on the many facets of quality.

From a practical perspective, one of the main issues is to comply with the requirement of relevancy of digital evidence. Indeed, in Digital Forensics, *a priori* discarding of data is not recommended, because of the risk of deleting important details. Yet Forensics experts need to understand which type of data must be acquired and how it should be manipulated. From this view, IQ could be considered a kind of metadata: information *about* information. In a nutshell, to analyse IoT evidence, it is essential to track data from the original source to the data repository. IoT Forensics simply follows the path of information going backwards.

IoT Forensics is one of the new subareas of digital Forensics [77, 37], and very few studies have been carried out with the specific aim of modelling theoretical approaches [55, 15, 5].

In this contribution, we drew a theoretical framework and introduced a model for assessing IQ in IoT Forensics. Our approach, which we expressed in a general formula and three applications — one for each layer — aims to describe the complexity of IQ in the IoT environment. To the best of our knowledge, this is the first attempt to provide a mathematical framework for a purely theoretical issue, that of the legal argumentation for evaluating evidence in trial, in the case of the IoT environment.[33] Further steps more steps need to be made along this path.[34]

---

[33] This involves technical evaluation of the quality of a new kind of evidence, which could be acquired from different and heterogeneous sources, maybe without a law enforcement agent being physically on site.

[34] After a previous introduction [19], we made a further step [31] but many more are required.

First, we aim to implement our approach, exploring the possibility of creating a database for every data type, thus allowing us to obtain robust statistics, and in so doing, developing an approach which involves Gaussian mixture models theory [62, 65, 70].

Second, in the future we aim to implement our model integrating the ISO/OSI level of complexity by adding four more layers, based on the interaction with the human factor, which is pivotal in Forensics, since the evidence has to be discussed in courts.

Third, we aim to provide further details on the features of the observer, now generally expressed as OS.

Fourth, it is important to assess the impact of this approach on a legal aspect, evaluating if the rights and fundamental freedoms of individuals — indicted and victims — could be endangered by its implementation. It is obvious that most threats could emerge in the processing of personal data, thus posing risks for discrimination and unfair treatment.

Fifth, we believe that benchmarks should be established in order to assess the reliability of our model. To create such benchmarks, the possibility of including data from previous evaluations could be explored. For example, one could reuse data concerning the analysis of certain models of devices (i.e. the Samsung "smart tv" model UE65NU7400U) in order to fine-tune certain variables (i.e. accessibility) and, in a more sophisticated model, integrate such assessment with technical specifications provided directly by manufacturers and feedback from the community of forensic experts. Perhaps in a relatively near future devices could be classified not only in terms of energy consumption but also by their cybersecurity or forensic trustworthiness.

References

[1] Amina Adadi and Mohammed Berrada. Peeking Inside the Black-Box: A Survey on Explainable Artificial Intelligence (XAI). *IEEE Access*, 6:52138–52160, 2018.

[2] Albert J. Ahumada. Computational Image Quality Metrics: A review. *SID Digest*, 24:305–308, 1993.

[3] Matthew P. J. Ashby. The Value of CCTV Surveillance Cameras as an Investigative Tool: An Empirical Analysis. *European Journal on Criminal Policy and Research*, 23(3):441–459, 2017.

[4] Kevin Ashton. *That 'Internet of Things' Thing*. 2009. Publication Title: RFID Journal.

[5] Sachin Babar, Parikshit Mahalle, Antonietta Stango, Neeli Prasad, and Ramjee Prasad. Proposed Security Model and Threat Taxonomy for the Internet of Things (IoT). In Natarajan Meghanathan, Selma Boumerdassi, Nabendu Chaki, and Dhinaharan Naga-

malai, editors, *Recent Trends in Network Security and Applications*, pages 420–429. Springer, Berlin, Heidelberg, 2010.

[6] Yehoshua Bar-Hillel and Rudolf Carnap. Semantic Information. *The British Journal for the Philosophy of Science*, 4(14):147–157, 1953.

[7] Carlo Batini, Cinzia Cappiello, Chiara Francalanci, and Andrea Maurino. Methodologies for Data Quality Assessment and Improvement. *ACM Computing Surveys*, 41(3):1–52, 2009.

[8] Carlo Batini and Monica Scannapieco. *Data and Information Quality: Dimensions, Principles and Techniques*. Data-Centric Systems and Applications. Springer International Publishing, 2016.

[9] Sebastiano Battiato, Oliver Giudice, and Antonino Paratore. Multimedia Forensics: Discovering the History of Multimedia Contents. In *Proceedings of the 17th International Conference on Computer Systems and Technologies 2016*, CompSysTech '16, pages 5–16, New York, NY, USA, 2016. ACM.

[10] Majid Davoody Beni. Syntactical Informational Structural Realism. *Minds and Machines*, 28(4):623–643, 2018.

[11] Albert Borgmann. *Holding on to Reality. The Nature of Information at the Turn of the Millennium*. University of Chicago Press, Chicago, 1999.

[12] Matthew Bovee, Rajendra P. Srivastava, and Brenda Mak. A Conceptual Framework and Belief-function Approach to Assessing Overall Information Quality. *International Journal of Intelligent Systems*, 18(1):51–74, 2003.

[13] A. Boztas, A. R. J. Riethoven, and M. Roeloffs. Smart TV forensics: Digital traces on televisions. *Digital Investigation*, 12:S72–S80, 2015.

[14] Sotirios Brotsis, Nicholas Kolokotronis, Konstantinos Limniotis, Stavros Shiaeles, Dimitris Kavallieros, Emanuele Bellini, and Clement Pavue. Blockchain Solutions for Forensic Evidence Preservation in IoT Environments. In *2019 IEEE Conference on Network Softwarization (IEEE NetSoft), , Paris, France*, Paris, 2019.

[15] Dong Chen, Guiran Chang, Dawei Sun, Jiajia Li, Jie Jia, and Xingwei Wang. TRM-IoT: A Trust Management Model Based on Fuzzy Reputation for Internet of Things. *Computer Science and Information Systems*, 8(4):1207–1228, 2011.

[16] Devon R. Clark, Christopher Meffert, Ibrahim Baggili, and Frank Breitinger. DROP (DRone Open source Parser) Your drone: Forensic Analysis of the DJI Phantom III. *Digital Investigation*, 22:S3–S14, 2017.

[17] Louis Columbus. *2018 Roundup Of Internet Of Things Forecasts And Market Estimates*. 2018. Publication Title: Forbes.

[18] Mauro Conti, Ali Dehghantanha, Katrin Franke, and Steve Watson. Internet of Things Security and Forensics: Challenges and Opportunities. *Future Generation Computer Systems*, 78:544–546, 2018.

[19] Federico Costantini, Marco Alvise De Stefani, and Fausto Galvan. The ÂńQuality of InformationÂż Challenges in IoT Forensics: An Introduction. *Jusletter IT*, (21 February 2019), 2019.

[20] Philip B. Crosby. *Quality is Free: the Art of Making Quality Certain.* McGraw-Hill, New York, 1979.

[21] William H. DeLone and Ephraim R. McLean. Information Systems Success: The Quest for the Dependent Variable. *Information Systems Research,* 3(1):60–95, 1992.

[22] William H. DeLone and Ephraim R. McLean. The DeLone and McLean Model of Information Systems Success: A Ten-Year Update. *Journal of Management Information Systems,* 19(4):9–30, 2003.

[23] Fred I. Dretske. *Knowledge & the Flow of Information.* MIT Press, Cambridge, Mass., 1981.

[24] Peter G. Engeldrum. Psychometric scaling: avoiding the pitfalls and hazards. In *PICS,* pages 101–107, 2001.

[25] Luciano Floridi. *The Philosophy of Information.* Oxford University Press, Oxford, 2013.

[26] Luciano Floridi. *The Onlife Manifesto. Being Human in a Hyperconnected Era.* Open Access. Springer International Publishing, Cham, 2015.

[27] Luciano Floridi. *The 4th Revolution: How the Infosphere is Reshaping Human Reality.* Oxford University Press, Oxford, 2016.

[28] Luciano Floridi. *The Logic of Information: A Theory of Philosophy as Conceptual Design.* 2019.

[29] Luciano Floridi and Phyllis Illari. *The Philosophy of Information Quality.* Synthese library. Springer, Berlin-Heidelberg, 2014.

[30] Michael Friedewald and Oliver Raabe. Ubiquitous Computing: An Overview of Technology Impacts. *Telematics and Informatics,* 28(2):55–65, 2011.

[31] Fausto Galvan, Federico Costantini, and Sebastiano Battiato. A Case Study for an "Accountable" IoT Forensics. In Eric Schweighofer, Walter Hötzendorfer, Franz Kummer, and Ahti Saarenpää, editors, *Verantwortungsbewusste Digitalisierung / Responsible Digitalization. Tagungsband des 23. Internationalen Rechtsinformatik Symposions IRIS 2020 / Proceedings of the 23rd International Legal Informatics Symposium IRIS 2020,* Colloquium, pages 533–542. Weblaw, Bern, 2020.

[32] Fausto Galvan, Giovanni Puglisi, Arcangelo Ranieri Bruna, and Sebastiano Battiato. First Quantization Matrix Estimation From Double Compressed JPEG Images. *IEEE Transactions on Information Forensics and Security,* 9(8):1299–1310, 2014.

[33] Anthony Giddens. *The Constitution of Society.* University of California Press, Berkeley, 1984.

[34] Uwe Hansmann. *Pervasive Computing Handbook.* Springer, Berlin; New York, 2001.

[35] Robert Hegarty, David J. Lamb, and Andrew Attwood. Digital Evidence Challenges in the Internet of Things. In *Proceedings of the Tenth International Network Conference (INC) 2014,* pages 163–172. School of Computing & Mathematics Plymouth University, Plymouth, 2014.

[36] M. Hossain, Y. Karim, and R. Hasan. FIF-IoT: A Forensic Investigation Framework for IoT Using a Public Digital Ledger. In *2018 IEEE International Congress on Internet*

of Things (ICIOT), pages 33–40. IEEE, 2018.

[37] Jan Höller, Vlasios Tsiatsis, Catherine Mulligan, Stamatis Karnouskos, Stefan Avesand, and David Boyle. *From Machine-To-Machine to the Internet of Things: Introduction to a New Age of Intelligence.* Academic Press, Amsterdam, 2014.

[38] ISO. ISO 9000:2015 Quality management systems — Fundamentals and vocabulary, 2015.

[39] ISO/IEC. 25012:2008 Software engineering — Software product Quality Requirements and Evaluation (SQuaRE) — Data quality model, 2008.

[40] ISO/IEC. ISO/IEC 27037:2012 Information technology — Security techniques — Guidelines for identification, collection, acquisition and preservation of digital evidence, 2012.

[41] ISO/IEC. ISO/IEC 27041:2015 Information technology — Security techniques — Guidance on assuring suitability and adequacy of incident investigative method, 2015.

[42] ISO/IEC. ISO/IEC 27042:2015 Information technology — Security techniques — Guidelines for the analysis and interpretation of digital evidence, 2015.

[43] ISO/IEC. ISO/IEC 27050-1:2016 Information technology — Security techniques — Electronic discovery — Part 1: Overview and concepts, 2016.

[44] ISO/TS. ISO/TS 8000-1:2011 Data quality — Part 1: Overview, 2011.

[45] ITU-T SG20. Recommendation ITU-T Y.2060 Overview of the Internet of Things, 2012.

[46] Matthias Jarke, Maurizio Lenzerini, Yannis Vassiliou, and Panos Vassiliadis. *Fundamentals of Data Warehouses.* Springer-Verlag, Berlin Heidelberg, 2 edition, 2003.

[47] J. M Juran. *Juran on Planning for Quality.* Free Press ; Collier Macmillan, New York; London, 1988.

[48] Aimad Karkouch, Hajar Mousannif, Hassan Al Moatassime, and Thomas Noel. Data quality in Internet of Things: A State-of-the-Art Survey. *Journal of Network and Computer Applications*, 73:57–81, 2016.

[49] Ralf Küsters, Tomasz Truderung, and Andreas Vogt. Accountability: Definition and Relationship to Verifiability. pages 526–535, New York, NY, USA, April 2010. ACM. Backup Publisher: Proceedings of the 17th ACM conference on Computer and communications security.

[50] Liu Liping and Lauren N. Chi. Evolutional Data Quality. In *7th International Conference on Information Quality*, pages 292–304. MIT, Cambridge, 2002.

[51] Björn Lundgren. Does Semantic Information Need to be Truthful? *Synthese*, 196:2885–2906, 2017.

[52] Andrzej Magruk. The Most Important Aspects of Uncertainty in the Internet of Things Field — Context of Smart Buildings. *Procedia Engineering*, 122:220–227, 2015.

[53] Massimo Mancuso and Sebastiano Battiato. An Introduction to the Digital Still Camera Technology. *ST Journal of System Research*, 2(2):1–9, 2001.

[54] Christopher Meffert, Devon Clark, Ibrahim Baggili, and Frank Breitinger. Forensic State Acquisition from Internet of Things (FSAIoT): A General Framework and Prac-

tical Approach for IoT Forensics Through IoT Device State Acquisition. In *Proceedings of the 12th International Conference on Availability, Reliability and Security*, pages 1–11. ACM, Reggio Calabria, Italy, 2017.

[55] Natasha Micic, Daniel Neagu, Felician Campean, and Esmaeil Habib Zadeh. Towards a Data Quality Framework for Heterogeneous Data. pages 155–162, 2017.

[56] Felix Naumann. *Quality-Driven Query Answering for Integrated Information Systems*. Lecture Notes in Computer Science. Springer-Verlag, Berlin Heidelberg, 2002.

[57] Nicole R. Odom, Jesse M. Lindmar, John Hirt, and Josh Brunty. Forensic Inspection of Sensitive User Data and Artifacts from Smartwatch Wearable Devices. *Journal of Forensic Sciences*, 2019.

[58] Gary Palmer. A Road Map for Digital Forensic Research. Report From the First Digital Forensic Research Workshop (DFRWS), 2001.

[59] Digambar Povar and G. Geethakumari. A Heuristic Model for Performing Digital Forensics in Cloud Computing Environment. In JaimeLloret Mauri, SabuM Thampi, DandaB Rawat, and Di Jin, editors, *Security in Computing and Communications*, volume 467 of *Communications in Computer and Information Science*, pages 341–352. Springer Berlin Heidelberg, 2014.

[60] Judith A. Redi, Wiem Taktak, and Jean-Luc Dugelay. Digital Image Forensics: A Booklet for Beginners. *Multimedia Tools and Applications*, 51(1):133–162, 2011.

[61] Thomas C. Redman. *Data Quality for the Information Age*. Artech House, Inc., Norwood, MA, USA, 1st edition, 1997.

[62] Douglas Reynolds. Gaussian Mixture Models. In Stan Z. Li and Anil Jain, editors, *Encyclopedia of Biometrics*, pages 659–663. Springer US, Boston, MA, 2009.

[63] de Huib Ridder and Serguei Endrikhovski. 33.1: Invited Paper: Image Quality is FUN: Reflections on Fidelity, Usefulness and Naturalness. *SID Symposium Digest of Technical Papers*, 33(1):986–989, 2002.

[64] John Roberts, Terry McNulty, and Philip Stiles. Beyond Agency Conceptions of the Work of the Non-Executive Director: Creating Accountability in the Boardroom. *British Journal of Management*, 16(s1):S5–S26, 2005.

[65] Phil Rose and Elaine M. Winter. Traditional Forensic Voice Comparison with Female Formants: Gaussian Mixture Model and Multivariate Likelihood Ratio Analyses. pages 42–45, 2010.

[66] Eliyahu Safra, Yaron Kanza, Yehoshua Sagiv, Catriel Beeri, and Yerach Doytsher. Location-based Algorithms for Finding Sets of Corresponding Objects over Several Geo-spatial Data Sets. *International Journal of Geographical Information Science*, 24(1):69–106, January 2010.

[67] Palaniappan Shamala, Rabiah Ahmad, Ali Zolait, and Muliati Sedek. Integrating Information Quality Dimensions into Information Security Risk Management (ISRM). *Journal of Information Security and Applications*, 36:1–10, 2017.

[68] Stavros Simou, Christos Kalloniatis, Evangelia Kavakli, and Stefanos Gritzalis. Cloud Forensics: Identifying the Major Issues and Challenges. In Matthias Jarke, John My-

lopoulos, Christoph Quix, Colette Rolland, Yannis Manolopoulos, Haralambos Moura-tidis, and Jennifer Horkoff, editors, *Advanced Information Systems Engineering*, volume 8484 of *Lecture Notes in Computer Science*, pages 271–284. Springer International Publishing, 2014.

[69] Richard Y. Wang and Diane M. Strong. Beyond Accuracy: What Data Quality Means to Data Consumers. *Journal of Management Information Systems*, 12(4):5–33, 1996.

[70] Xinyi Wang, Shaozhang Niu, and Jiwei Zhang. Digital Image Forensics Based on CFA Interpolation Feature and Gaussian Mixture Model. *IJDCF*, 11(2):1–12, 2019.

[71] S. Wayne. Quality Control Circle and Company Wide Quality Control. *Quality Progress*, October:14–17, 1983.

[72] Warren Weaver. The Mathematics of Communication. *Scientific American*, 181(1):11–15, 1949.

[73] Mark Weiser. The Computer for the 21st Century. *SIGMOBILE Mob. Comput. Commun. Rev.*, 3(3):3–11, 1999.

[74] Gary B. Wills, Ahmed Alenezi, Nik Zulkipli, and Nurul Huda. IoT Forensic: Bridging the Challenges in Digital Forensic and the Internet of Things. pages 315–324. Scitepress, 2017.

[75] Philip Woodall, Alexander Borek, and Ajith Kumar Parlikad. Data quality assessment: The Hybrid Approach. *Information & Management*, 50(7):369–382, 2013.

[76] Zheng Yan, Peng Zhang, and Athanasios V. Vasilakos. A Survey on Trust Management for Internet of Things. *Journal of Network and Computer Applications*, 42:120–134, June 2014.

[77] Ibrar Yaqoob, Ibrahim Abaker Targio Hashem, Arif Ahmed, S. M. Ahsan Kazmi, and Choong Seon Hong. Internet of Things Forensics: Recent Advances, Taxonomy, Requirements, and Open Challenges. *Future Generation Computer Systems*, 92:265–275, 2019.

[78] Sergej Yendrikhovskij. Image quality: Between science and fiction. In *PICS*, pages 173–178, 1999.

[79] Muhammad Sharjeel Zareen, Adeela Waqar, and Baber Aslam. Digital Forensics: Latest Challenges and Response. In *2013 2nd National Conference on Information Assurance (NCIA)*, pages 21–29, Piscataway, NJ, 2013. IEEE.

[80] S. Zawoad and R. Hasan. FAIoT: Towards Building a Forensics Aware Eco System for the Internet of Things. In *2015 IEEE International Conference on Services Computing*, pages 279–284. IEEE, July 2015.

 Received 23 May 2019

DIFFERENT WAYS OF BEING NAKED. A SCENARIO APPROACH TO THE NAKED STATISTICAL EVIDENCE PROBLEM

ANNE RUTH MACKOR
University of Groningen
`a.r.mackor@rug.nl`

1 Introduction

Suppose that it is known with sheer certainty that there were 100 persons on an island, that 99 of them were involved in a crime and that no other information is available. Is it legitimate, given the high probability that they were involved, to convict one, some or all of these persons? I surmise that the answer of most people will be a straightforward "no". Without any 'specific' or 'individualized' evidence it seems wrong to conclude that the criminal facts have been proven beyond a reasonable doubt. This, in a nutshell, is the problem of naked statistical evidence.

It has turned out to be very hard to argue exactly why judicial decisions that a crime (or a tort) has been proven should not be based only on 'naked' statistical evidence. Despite the fact that the naked statistical evidence problem has been discussed for more than forty years now,[1] philosophers, legal scholars and probability theorists still offer divergent and sometimes conflicting arguments why a high probability on its own should not suffice to decide a case against a defendant.

The purpose of this article is different from other articles on the topic. I will not give an overview of the most important epistemological and moral arguments that have been offered thus far, nor will I present my own arguments against the use of naked statistical evidence. The focus of this article is on a preliminary matter, viz

---

I thank two anonymous reviewers, Christian Dahlman, Marcello Di Bello and Ronald Meester for their comments on an earlier version of my paper. I thank the participants of the North Sea Evidence Group of Legal Evidence and Proof Workshop on Bayes and Inference to the Best Explanation in Criminal Law, Groningen, November 2019, for their comments on my talk on the topic.

[1]For overviews of arguments see for example [25]; [13].

the nature of naked statistical evidence. I analyse an aspect of the problem that has not been discussed in the literature so far, namely that there are different ways of being naked.

My article is structured as follows. In section 2, I present four legal cases that are often used to illustrate the naked statistical evidence problem. I present two civil cases, but the focus of my article is on criminal law. In section 3 I introduce the scenario approach to reasoning about evidence and proof in criminal law. Subsequently, in section 4, I distinguish between evidence that can be causally explained by a scenario (trace evidence), and other information that cannot be causally explained but that seems to be relevant for the case at hand (non-trace evidence). Next, in section 5 I distinguish different kinds of trace evidence. In section 6 I apply these distinctions to the four cases and argue for my main claim, viz that there are different ways of being naked. I show that two of the cases focus on non-trace evidence (section 7), whereas the other two focus on trace evidence (section 8) and argue that the two types of naked statistical evidence problem ask for different analyses. I conclude that my analysis of the different ways of being naked contributes to clarifying the nature of naked statistical evidence (section 9).

2   Four legal cases

In this section I describe four legal cases that are often used in the literature on the naked statistical evidence problem. Two are civil cases, two are criminal cases; two are hypothetical, one is based on real legal case and the last is a real legal case.[2]

The Gatecrasher case [3]
The first hypothetical case is a civil case about a rodeo show. In a variant of the case, there is ample evidence that the rodeo show has been attended by 1000 visitors and there is also evidence that only 499 tickets have been sold. From these findings it is concluded that 501 visitors have crashed the gate. The organizer of the show starts a law suit against the visitors. It turns out that none of visitors are able to present evidence that they paid for their tickets. Thorough investigations have been carried out, but they have not yielded any further evidence. The probability that further evidence could still come to light is negligible. The evidence about the number of visitors and the number of gate crashers is not disputed. So, all other things being equal, the probability that a visitor crashed the gate is deemed to be

---

[2]There are many different versions of the first three cases. In this section I offer my own variants of the cases.
   [3][4].

slightly more than 50%.

Although the explication of standards of proof in terms of probabilities has been criticized, the standard of preponderance of evidence has been equated to a probability of more than 50%.[4] Accordingly, the question to which the answers of philosophers, probability theorists and legal scholars diverge is whether and why the claimant is or is not entitled to judgment against one, some or even all of the defendants in view of the fact that the probability that each individual defendant crashed the gate is more than 50%.

The Prison Riot or Prison Yard case[5]
A criminal variant of the Gatecrasher case goes as follows. One hundred prisoners, whose identities are known, were in the courtyard of a prison, guarded by one guard. At a certain moment the prisoners, who had devised a plan to escape, attacked the guard and killed him. The images from the security camera show that all but one of the prisoners have participated in the attack. However, the camera images are too coarse-grained to identify the prisoners. Apart from the injuries on the corpse of the guard, the camera images are the only evidence about the riot and the killing; there are no witnesses and there is no 'technical' evidence such as DNA or fingerprints. Thorough investigations have been carried out, but they have not yielded any further evidence. The probability that other evidence could still come to light is therefore negligible. When the case is brought to court, all defendants claim to be the one innocent prisoner.

Again, although the explication of standards of proof in terms of probabilities has been criticized, the threshold of beyond a reasonable doubt is sometimes equated with a probability of at least 95% or 90%.[6] With a probability of 99%, this threshold is amply reached in this case. Again, the question is: could or even should the jury or judge consider the criminal facts proven beyond a reasonable doubt on the basis of this 'naked' probability? And thus, could or should one, some, or all of the hundred prisoners be convicted?

---

[4]See for example [3] at 1256, note 1 and 2 for references to probabilistic explications of the standards of proof. Cheng states that the preponderance standard is conventionally described as an absolute probability threshold of 0.5, but argues that the standard should be reconceptualized as a probability ratio. More on Cheng's view in section 8. Also see for example [1] who argue that standards of proof are not probabilistic but explanatory thresholds.

[5][22].

[6][3], 1256, note 2. See above, note 4.

The Blue Bus case[7]

The third case, which is again a civil case, is based on the American lawsuit Smith v. Rapid Transit.[8] In the real case, a bus hit the car of Betty Smith around one o'clock at night. The bus did not stop after the collision. There were no witnesses apart from Betty Smith who had only seen that her car had been hit by a bus. Accordingly, this might seem like a court case - against whom? - without a chance. Smith, however, started a lawsuit against the local bus company Rapid Transit. In court, she presented evidence that this company provided a scheduled service via the road on which Smith was hit by a bus and the timetable showed that buses of this company actually drove via this road at night around the time of the collision. However, the Supreme Court of Massachusetts rejected the claim, arguing, among others, that it is "not enough that mathematically the chances somewhat favour a proposition to be proved."

One version of the hypothetical Blue Bus case goes as follows. Again, Betty Smith's car is hit by a bus and again she does not recall anything else than that her car has been hit by a bus. She starts a lawsuit against Blue Bus company. She presents evidence that all buses in town are operated by two bus companies, Blue Bus and Red Bus and that Blue Bus has a market share of 80%. In this case too, after thorough investigation, there is no further evidence. The claim that Smith's car was hit by a bus from Blue Bus company is based on the market share of 80%.[9] As in the other cases, the question arises whether Betty Smith is entitled to a judgement against Blue Bus on the basis of this naked statistical evidence.

The case of Shonubi[10]

The criminal case United States v. Shonubi is, like Smith v. Rapid Transit, a real American case. Shonubi was convicted for drug smuggling after having been caught at JFK and found to have swallowed 103 balloons filled with 427.4 grams of heroin in total. Up to this point, it was a 'normal' case with sufficient 'individualized' evidence; Shonubi could be convicted for transporting 427.4 grams of heroin.

However, a dispute arose about the total quantity of drugs Shonubi had smug-

---

[7][28].

[8]Smith v. Rapid Transit, 317 Mass. 469, 58 N.E.2d 754 (1945). For a more extensive description and a discussion of Smith v. Rapid Transit and the Blue Bus problem, see for example [26], 79 ff. See [27] for an analysis of another hypothetical variant of this case, Smith v. Red Cab.

[9]The probability in the Blue Bus case not only seems to meet the standard of preponderance of evidence, but also that of clear and convincing evidence, a standard that has been equated to a probability of more than 75%.

[10]I present a simplified version of the case. I have relied on the description of [16] and on the description and the analysis of [5]. Also see [6].

gled. The total quantity was important because the maximum sentence for drug smuggling varied with the total quantity of drugs being transported. Further investigations had produced undisputed evidence that Shonubi had travelled between Nigeria and JFK on seven earlier occasions and that the purpose of these trips had been drug smuggling too. Therefore, the prosecution had an interest in proving that the total quantity that Shonubi transported was much more than these 427.4 grams. The prosecution tried to provide evidence for the quantity of drugs he had smuggled on his previous trips and take that to be the total quantity.

Apart from the fact that this interpretation of "total" seems to stretch its meaning far beyond the normal interpretation of 'total on one trip', it also results in an interesting version of the naked statistical evidence problem. In first instance, the prosecution claimed that it was very likely that Shonubi had transported the same quantity of heroin on each of the eight trips he made, and thus that he had transported 3419.2 grams in total. The lower court sentenced him in accordance with this estimate. Shonubi, however, appealed, claiming that this estimate was "speculative" and the appellate court agreed that any quantity beyond 427.4 grams was speculative. It stated that the prosecution had not produced "specific evidence" about the total quantity, for example a testimony or drugs reports, and sent the case back to the district court.

This time, the prosecution delivered a lengthy report containing information about the gross weight of heroin smuggled by 117 Nigerian drug smugglers who had been caught at JFK in the same period that Shonubi travelled between Nigeria and JFK. Based on this information, it was estimated that the probability that Shonubi transported at least 2090.2 grams on these seven trips was 99%. Again, the lower court accepted the evidence and sentenced him for transporting 1000 to 3000 grams of heroin. Shonubi however, appealed again and the appellate court repeated its demand for "specific evidence" and denied that the data about other drug smugglers was the specific evidence needed. In the end, the lower court sentenced Shonubi on the basis of the quantity of heroin he had transported on his eighth trip, viz 427.4 grams.

'Individualized' evidence versions of the four cases
In all four cases, naked statistical evidence plays a pivotal role. To clarify that role, let us first contrast the 'naked' cases with versions in which there is 'individualized' or 'specific' evidence. For example, statements of reliable eyewitnesses who state that they saw a particular defendant crash the gate or that they saw a particular defendant hit the prison guard are 'individualized' evidence that can, in principle,

be sufficient for a decision against the defendant.[11] The same holds for the individualized version of the Blue Bus case in which Betty or any other reliable witness does not just state that she saw a bus, but that she saw a blue bus. Finally, in the Shonubi case, the custom officers might state that they had caught Shonubi with a specific quantity of heroin on each of the seven earlier trips.

One of the individualized variants of the Blue Bus case in the literature goes as follows. In this variant, Blue Bus bus company does not own 80%, but only 50% of the buses; Red Bus owns the other 50%. Second, there is some individualized evidence, viz a witness stating that she saw that the car was hit by a blue bus. The witness is reliable, but not infallible: experiments carried out under the same nightly conditions have shown that she correctly identifies the colour of the bus in 80% of the cases. In other words, in 80% of the cases in which there actually is a blue bus, the witness states that she sees a blue bus.[12] Thus, the probability that Smith was hit by a blue bus seems to be the same as in the naked statistical evidence variant.

Moreover, the example shows that there is no such thing as individualized evidence in which statistics do not play a role. On the contrary, all individualized evidence relies on statistics, in this particular case on statistics about the reliability of the witness. Nevertheless, many persons who would deny Betty Smith's claim in the naked Blue Bus case would be willing to order Blue Bus to pay the damages in the individualized version, even though her testimony is fallible and even though the assessment of her testimony is based on 'underlying' statistics. Accordingly, the distinction between 'naked' statistics and 'underlying' statistics seems to be relevant for legal decisions.

Most authors present cases of naked statistical evidence as if they are all of the same kind. Below, in sections 6 to 8, I will argue that the nakedness in the Gatecrasher and Prison Yard cases is different from the nakedness in the Blue Bus and Shonubi cases. Before I can do so, I must first introduce the scenario approach to evidence and proof (section 3) in which the distinction between trace evidence and non-trace evidence (section 4) and distinctions between different types of trace

---

[11] More fine-grained camera images that allow one to identify the different prisoners in the yard could do the same job as eyewitnesses who testify that they identified the prisoner who hid away. They might even do a better job, since camera's do not suffer from memory problems and their images can be checked more than once and by different persons.

[12] The reliability of the observations of the witness is not the only factor that must be taken into account. Among others, the memory, the cognitive capacities and the honesty of witnesses are not infallible either.

evidence (section 5) play a pivotal role. In sections 6 to 8 I apply these distinctions to the four examples of naked statistical evidence to show how their nakedness differs.

3 A scenario approach to evidence and proof in criminal cases

The scenario approach builds on the story model of Pennington and Hastie which describes how people reason when they have to deal with a lot of evidence about a case. Crombag, Van Koppen and Wagenaar have turned the story model into a normative model about how people should reason if they want to make rational or reasonable decisions.[13] I will briefly describe five main characteristics of this normative scenario approach.

To illustrate the scenario approach, I will use a simplified version of a Dutch criminal case, the Simonshaven case.[14] Ed, the defendant, and Jenny, his wife, arrived at the Simonshaven forest by car. They went for a walk. There, Jenny was brutally attacked and died. There are two scenarios. According to the prosecution scenario Ed killed his wife. On the defense scenario a madman jumped out of the bushes who beat up both Ed and Jenny as a consequence of which Ed temporarily lost his consciousness and Jenny died.

1. A scenario consists of elements
A first thing to note is that a scenario is more than a single hypothesis about the criminal act. A scenario is a story which consists of different elements, viz initiating events, a psychological response, sometimes a goal, an action, one or more consequences. All elements are hypotheses about what happened. Together they make up a story.

2. Elements are chronologically ordered and causally connected
A second important characteristic is that scenarios have a specific structure; the elements are chronologically ordered and they are connected through physical or mental causal relationships. Among others, the psychological state and the motive are causally connected to the act and the act (beating) is causally connected to the

---

[13] [24]. [32]; [30]. My description derives from the most recent exposition of the scenario approach in [31].

[14] The Dutch Simonshaven case is analyzed more extensively in [31]. For the decision of the Court of Appeal of The Hague, February 18 2015, ECLI:NL:HR:2016:2201, see https://uitspraken. rechtspraak.nl/inziendocument?id=ECLI:NL:HR:2016:2201 (last visited July 24, 2019).

consequences (injuries, death).

3. Inference to the best explanation

Until now I explicated the internal structure of scenarios. The next question is how scenarios and evidence relate. The scenario approach is a species of a well-known general approach to evidence and proof, i.e. inference to the best explanation. Inference to the best explanation instructs factfinders to assess and compare and different explanations and the best explanation is the one that offers the best explanation of those facts.[15] Similarly, on the scenario approach fact finders should assess and compare scenarios and find out which scenario offers the best explanation of the evidence in the case at hand.

4. General background knowledge, internal coherence and evidence

When determining which scenario offers the best explanation, fact finders should assess and compare the explanatory relation between the scenarios and the evidence in a case. However, they should also assess and compare the internal coherence of the scenarios (internal consistency, story gaps, level of detail, etc.). The defense scenario in the Simonshaven case lacks detail and has story gaps. Among others there is a story gap between the death of Jenny and the first time Ed makes a phone call to one of his children. Fact finders should also assess the fit of the scenarios with general background knowledge (their general plausibility). Clearly, a 'madman scenario' fits less well with our general assumptions about the world than a 'husband kills wife scenario'.

5. Causal explanations of the evidence

We have seen that the relations between the elements within the scenario are chronologically ordered (Ed and Jenny arrived at the forest, then they went for a walk) and that some elements are causally connected (Ed beat up Jenny, this caused her death). Next to that, the relations between the elements of the scenario and the evidence are also causal. That is to say, the explanations that the scenario approach demands are causal explanations of the evidence about the case at hand and the best scenario should offer better explanations than any of the other scenarios. So, for example, the prosecution scenario in the Simonshaven case hypothesizes that Ed hit Jenny with a blunt object, possible the butt of a gun. This hypothesis can causally explain the evidence about Jenny's injuries. However, the defense scenario can also causally explain the evidence about the injuries by hypothesizing the madman caused them.

---

[15]See for instance [15, 19] and [12].

Note that although the best scenario should be able to causally explain the evidence, the evidence need not actually have been caused by a state of affairs, actions and events hypothesized by the scenario. The only demand is that it is possible that the evidence has been caused in the manner explained by the scenario. The scenario approach does not exclude the possibility that the evidence has been caused in another way. In other words, the best causal explanation need not be the true explanation and incorrect convictions and acquittals are possible.

In order to reach the threshold of beyond a reasonable doubt in criminal law, the prosecution scenario should offer a much better causal explanation of the evidence than the defense scenario. This claim must be nuanced, however. Above, I pointed out that factfinders should also assess the internal coherence of the scenario and the fit of the scenario with our general beliefs about the world. Accordingly, the capacity of the scenario to causally explain the evidence in the case is only one aspect of the assessment of a scenario. If the defense scenario can explain the evidence roughly equally well as the prosecution scenario but is internally incoherent and/or wildly implausible, the defendant can nevertheless be convicted.[16]

In fact, this is what happened in the Simonshaven case. The scenario of the prosecution can explain the evidence. It can explain that witnesses have seen a man and a woman in a car before they went for a walk and it can explain Ed's statement about the madman as a lie caused by his wish to cover up for his crime. However, the defense scenario can also explain the evidence and it can do so roughly equally well as the prosecution scenario. A crucial difference, however, is the internal coherence of the defense scenario and its fit with our general background knowledge. One of the prosecutors in the case stated that the defendant might just as well have claimed that men from Mars killed Jenny, i.e. that the madman scenario was extremely implausible.

Probability via a causal route
The instruction of the scenario approach that fact finders must assess a scenario in terms of its capacity to causally explain the evidence is relevant for the analysis of the naked statistical evidence problem. Most importantly, the scenario approach requires more from the relation between a scenario and evidence than delivering

---

[16] In its ruling of March 16 2010, ECLI:NL:HR:2010:BK3359 the Dutch Supreme Court has stated that if the defense scenario is extremely improbably, implausible or incredible, court can reject the defense scenario. See `https://uitspraken.rechtspraak.nl/inziendocument?id=ECLI:NL:HR:2010:BK3359` (last visited July 24 2019). This is so, even if it offers a better causal explanation of the evidence than the scenario of the prosecution. See [20] and [17] for analyses of this case.

a high probability; it demands that it delivers that probability via a causal route. In fact, several authors have argued that what makes evidence 'individualized' or 'specific' is causality.[17] On these views, evidence is 'individualized' if it can have been caused by the defendant committing the crime under investigation.

The scenario approach seems to fit well with causal views of what makes evidence 'individualized' or 'specific'. However, there are two important differences between the scenario approach and causal views on the naked statistical evidence problem. In the first place, the scenario approach does not simply demand factfinders to offer a causal explanation. In line with the demands of inference to the best explanation, it instructs fact finders to compare different scenarios and assess which scenario can best explain the evidence. A second distinctive feature of the scenario approach is that fact finders should not only explain evidence that can have been caused by the criminal act and its consequences, but also evidence that can have been caused by the initiating events, the psychological response and the goal that are hypothesized by the scenarios about the crime. For example, the witness statements that there were two persons in a dark Mercedes at the parking lot on August 11 2009 between 18.50 hours and 20.00 hours lot shortly before the crime took place cannot be caused by the crime because the event took place before the crime was committed. However, the hypothesized event of Ed and Jenny arriving by a blue Mercedes at the parking lot of the forest on August 11 2009 around 19 hours belongs, as part of the initiating events, to both scenarios and thus both scenarios can causally explain the witness statements.

4   Trace evidence versus non-trace evidence

Until now I have only referred to evidence that can be causally explained by a scenario. I have also briefly alluded to general background knowledge that is relevant for the assessment of the plausibility of the scenario. As regards the case of Ed and Jenny, I already mentioned that madmen jumping out of bushes and randomly killing people are thought to be extremely rare events, at least in the Netherlands. Other general background information which seems relevant to this case is that 52% of women who are killed, have been killed by their (ex)partners.[18] Neither pieces of information can be causally explained by the scenarios; the scenarios only offer a causal explanation of the killing of Jenny, they do not causally explain earlier killings. Interestingly, in the Simonshaven case there was information that shone a

---

[17]The first to explicitly do so was [27].

[18]See for example [23].

different light on the belief that madmen jumping out of bushes are extremely rare. A year after the death of Jenny, it turned out that a madman had been operating in the vicinity of the Simonshaven forest. He had killed at least two women and used the same modus operandi; he jumped out of the bushes and randomly killed his victims. This information too seems relevant for the case. It can change the assessment of the defense scenario, which was initially deemed extremely implausible. Again, however, these facts cannot be causally explained by either of the scenarios.

In most cases, there is much more information that cannot be causally explained by the actions, states of affairs and events that are hypothesized by the scenario, but that does seem to be relevant to the assessment of a case at hand. Take the Simonshaven case again. Ed and Jenny were married, but they had marital problems. Ed had used violence against Jenny on earlier occasions and he had threatened to kill her if she would leave him. Neither the defense scenario nor the prosecution scenario can causally explain these well-established and seemingly relevant facts about Ed. Here the causality at best runs in the opposite direction, viz from the evidence to the scenario, in particular to Ed's motive. These facts seem to present further evidence for and raise the probability of the scenario of the prosecution.

Following Uviller, I will call evidence that can be caused by a state of affairs, action or event which is hypothesized by a scenario, trace evidence. I distinguish it from other information that might be relevant for a case, but that cannot be caused by what the scenario hypothesizes to have happened. Uviller calls the latter predictive evidence.[19]

Uviller's term "trace" evidence and "predictive" evidence are aptly chosen. The term "trace evidence" brings to mind the trace an animal leaves when he moves through the forest and which helps a hunter to chase it. When the hunter detects a trace, he can explain that it has most likely been recently caused by a deer. Subsequently, he can use the trace to predict where the animal has gone. For example, the hunter might claim that it has gone to a nearby watering hole because most deer in this area are known to visit the watering hole at this time of the day. Or he might recognize the specific deer from its trace and know that this specific deer normally goes to the watering hole at this time of the day. His claim that the deer has most likely gone to the watering hole, is partly based on trace evidence, but for the largest

---

[19][29]. I take the reference to these terms from [8]. Since legal and in particular criminal cases are mainly about proof of past actions and events, one might prefer the term "retrodictive" evidence to emphasize the fact that we predict past events, not future events. More about prediction and retrodiction in [20].

part on his knowledge about past behaviour either of this deer in particular, or of deer in general, or on both. The example shows that the term profiling evidence, which has also been used in the literature, is also fitting since the hunter bases his predictions either on an individual profile of the particular deer or on a group profile of deer in general.[20] However, I will not use the term "predictive" or "profiling" evidence, but instead use the more neutral term "non-trace" evidence.

If the hunter goes to the watering hole, he will either acquire trace evidence that the deer is there, viz when he actually sees the deer, or that it has been there, viz when he finds traces of the deer. Whatever the outcome of his prediction, before the hunter goes to the watering hole, he can only say that it is highly probable the deer has gone there, that he is confident that the deer has gone there and that he has reason to believe that the deer has gone there. It is only when he finds trace evidence, namely traces at the watering hole or seeing the deer near the watering hole, that he can be said to know the deer is or has been there. Moreover, if he does not find any trace evidence, he can still say that he is confident that the deer has nevertheless gone to the watering hole, but again he cannot say that he knows the deer has been there, not even if the probability that the deer had gone to the watering hole is very high.[21]

Similarly, in the Simonshaven case, the statements of Ed, of the forensic pathologist and of the witnesses are trace evidence that can be caused by the states of affairs, actions and events hypothesized by one or both of the scenarios about the case. Other information, such as the evidence that Ed had marital problems,[22] the evidence that a madman had been operating in the vicinity, but also the information that 52% of women who are killed, are killed by their (ex)partners, are non-trace evidence that is at most predictive of the hypothesized crime.

---

[20] For example, [11] use the term profiling evidence. They do not distinguish between individual and group profiles, however. More on the distinction between individual and group profiles in section 6 and 7.

[21] My hunter example is a variant of the lottery paradox which deals with the question when I can be said to be confident, justified to believe or to know that I did not win a lottery in which I had only a very small chance to win it. Before the draw I have good reasons to believe that I will not win the lottery, but only after the draw, e.g. when I read the winning numbers in the newspaper, I can be said to know I did not win. See among others [21]; [2]; [18]. Also see [27].

[22] Some readers might want to argue that the marital problems are part of the scenario. However, they are not part of the main scenario. At most they can be part of a sub-scenario which is used to prove e.g. that Ed had both a motive and the means to kill Jenny. Still, they would not be trace evidence, but they could be predictive of the motive and means. See [31], 11, sub 6.

Non-trace evidence: investigation and pre-trial detention versus the proof of crime
Non-trace evidence can be particularly useful in the investigative phase of a criminal case. On the basis of his observation of one trace of the deer and his knowledge about places that deer in general tend to visit and/or knowledge about this specific deer in particular, the hunter can first go to the watering hole to find out whether his prediction was correct instead of randomly searching for the deer. In a similar manner, in the investigative phase, a factfinder, after having discovered that Jenny died an unnatural death and that Ed was with her at the time of the crime, will probably take into account the fact that 52% of killed women are killed by their (ex)-partner) and that Ed had threatened Jenny on earlier occasions, when he decides to search for trace evidence that could be caused by events hypothesized by the scenario that Ed killed his wife. Moreover, the non-trace information might be a valid reason to treat Ed as a potential suspect.[23]

Similarly, judicial decisions about pre-trial detention are mainly based on non-trace evidence. Again, the information about Ed's earlier violent behaviour, in combination with the trace evidence that Ed was with his wife at the time of the crime might not only raise a reasonable suspicion against him but it might also, in combination with an estimation of his risk of committing (further) crimes and his risk of flight, be a reason to put Ed in pre-trial detention. The decision about pre-trial detention cannot be based on trace evidence because it is, by definition, largely based on an estimate of Ed's future actions.

In this respect, decisions about pre-trial detention are different from decisions about the proof of a crime. Trace evidence is not yet available in the investigative phase and it is, by definition, absent in decisions about pre-trial detention. In the case of decisions about the proof of past crimes, on the other hand, there can be trace evidence caused by the crime.

When I discussed the hunter example, I already mentioned that when we only have non-trace evidence, we can say that we are confident and justified to believe that something has happened or will happen, but we cannot say that we know that something has happened or will happen.[24] If we can only predict whether something will happen in the future, as in the decision about pre-trial detention, or, when we have just started investigations into a crime, retrodict that something has happened,

---

[23]Note however, that the role of some types of non-trace evidence in the investigative phase has also been contested, for example as regards ethnic profiling. See for example [14].

[24]See above the hunter example and the references in note 21 to the literature about the lottery paradox.

it seems inevitable and therefore acceptable that we cannot (or not yet) have trace evidence and no knowledge in this sense of the word. However, in decisions about proof of crimes, where we could find trace evidence, knowledge could be possible. The question thus is whether we should demand knowledge based on trace evidence about the cause of a state of affairs (such as Jenny's death) or whether confidence based on non-trace evidence suffices.

One reason to demand trace evidence, at least in legal cases, is that it makes estimates about probabilities and thus decisions about the proof of facts more robust, and therewith safer, than decisions that are largely based on non-trace evidence. First, if there is only non-trace evidence, one has to consider that there can have been all kinds of intervening causes which can lead to a different course of events than the course of events that we predict on the basis of non-trace evidence.[25] One way to deal with this problem is to search for trace evidence that makes one course of events more likely than others. Second, absence of trace evidence suggests that either a 'perfect crime', i.e. a crime that left no traces, was committed or that investigations were sloppy, or both. In the Simonshaven case for example, the madman-scenario was not seriously investigated until long after the crime, viz only when it turned out that actually a madman had been operating in the area. A 'perfect crime' variant of the Simonshaven case could be a case in which the only trace evidence is direct non-individualized, viz that Jenny has disappeared and that the search history of Ed's computer reveals that he had recently used the search term 'perfect crime'; otherwise there is only non-trace evidence, e.g. that Ed has been violent against other persons in the past.

5 Different types of trace evidence

Above I have explicated that the scenario theory instructs fact finders not only to explain how evidence can be caused by the hypothesized crime, but also to causally explain evidence that is related to other elements of the scenario. Therefore, I distinguish between evidence that can be caused by the criminal act and evidence that can be caused by other events hypothesized by the scenario such as, for example, the initiating events. In this section I distinguish between individualized and non-individualized trace evidence and between direct and indirect trace evidence.

First, trace evidence can be about a specific defendant. An example is the testimony of a witness who states that he saw the defendant. I will call this kind

---

[25] Also see [7].

of evidence individualized and contrast it with evidence that is non-individualized. Second, trace evidence can be about the specific criminal act and its consequences. An example is the testimony of experts who investigate the technical evidence and state that the perpetrator, whoever it was, hit the victim with a blunt object and that this hitting was the cause of death. I will call this kind of evidence direct and contrast it with evidence that can be causally explained by an element of the scenario, but not by the criminal act. I will call the latter kind of evidence non-direct.

It is important to note that I do not use the distinction between direct and non-direct evidence in the well-known sense in which it, or rather the distinction between direct and circumstantial evidence, has long been used in the past. Until at least the nineteenth century, it has been argued that eyewitness statements are 'direct' evidence in that there would be no 'intermediate link' between the event and the evidence, e.g. between there actually being a blue bus and seeing a blue bus. Direct evidence was contrasted with technical evidence such as for example traces of blue paint on Betty Smith's car which would be indirect in that we need a causal generalization like 'If there are traces of blue paint on the damaged part of the car after an accident, they are most likely caused by the other vehicle involved in the accident'. It is now generally understood that eyewitness evidence also relies on generalizations like 'If witnesses state that they see a bus, then in most cases this statement is caused by the fact that there was a bus'. Accordingly, direct evidence in the classical sense of the term does not exist. All evidence is indirect in that it relies on some 'underlying' generalization. In this article, I use the term direct evidence in another sense, viz to contrast it with evidence that can be caused by an element of the scenario, but not by the crime.

Combining both distinctions, individualized versus non-individualized (i.e. whether or not the evidence can be caused by the presence or the action of a particular defendant) and direct versus indirect (i.e. whether or not the evidence can be caused by the criminal act), we can distinguish four kinds of trace evidence. I will explicate and illustrate the distinctions by applying them to the Prison Yard case. I will compare the scenario that prisoner D93 took part in the riot to the scenario according to which he was the one who hid away.

However, before I do so, it is important to note that the question whether evidence is individualized can only be answered in relation to the scenarios under investigation. There is no such thing as individualized evidence per se.[26] For exam-

---

[26] Also see below, section 8. For a similar view, see [9].

ple, whereas Betty Smith's statement that her car was hit by a bus is not sufficiently individualized when we compare the blue bus and the red bus scenario, the evidence can be sufficiently individualized when we compare the blue bus scenario to a truck scenario. Accordingly, it is only in relation to the specific scenarios that are presented in the analyses of the four cases, that evidence can be said to be individualized or not. Keeping this in mind, let us now distinguish different types of trace evidence.

- Individualized, but indirect trace evidence
  In the Prison Yard case, there is individualized but indirect evidence that 100 prisoners, of whom the identities are known, were in the yard when the crime was committed. Both scenarios start from the hypothesis that these were the initiating events and thus both can explain the evidence. Another example of individualized but indirect evidence would be a witness stating that he was in the prison yard shortly before the time of the crime and that he saw defendant D93 in the yard. Accordingly, such evidence would be about a particular defendant, but it would be indirect in that it would not be evidence that could be caused by the defendant actually taking part in the riot or the killing of the guard. However, the evidence could play a role in the proof of the criminal fact. For one thing, since the defendant was in the yard, he had the opportunity to take part in the riot and the killing.

- Non-individualized and indirect trace evidence
  For example, a witness statement that he saw the guard with four unknown and unidentified men shortly before the time of the crime would be non-individualized and indirect evidence. It is not clear what role this kind of evidence could play in decisions about proof of a crime, but the information could play an important role in the investigative phase; the police would probably want to find out who these persons were and to talk with them.

- Direct, but non-individualized trace evidence
  For example, technical evidence that the guard was beaten up, would be direct trace evidence about the crime, but not about the culprit. Camera images of 99 prisoners taking part in the riot and attacking the guard would also be an example of direct non-individualized evidence. This is evidence that could be causally explained by the prosecution scenario claiming that defendant D 93 was one of the 99 who took part in the riot. However, it seems that the defense scenario, according to which 99 prisoners were involved and D93 hid away, can also explain the evidence.[27]

---

[27]I say more about this in section 8.

- Direct and individualized trace evidence
  An example of direct and individualized trace evidence would be a witness stating that he saw defendant D93 beating the guard. The evidence would be both about the crime and about the specific defendant. The prosecution scenario that D93 beat the guard could causally explain this evidence and it could probably do so much better than the defense scenario that D93 was the one prisoner not taking part in the riot.[28] Thus, this kind of evidence seems to be capable of discriminating between the two scenarios under consideration.

Individualized versus non-individualized non-trace evidence
Before we turn to an analysis of the four cases, let us briefly consider whether we can also apply the distinction between individualized and non-individualized evidence to non-trace evidence. It seems that we can. For example, there is individualized but indirect non-trace evidence in the Blue Bus case and in the first version of the Shonubi case.

The information that the only buses in town are either blue or red and the information about the market share of Blue Bus is individualized albeit indirect non-trace evidence about Blue Bus. In the Shonubi case the individualized indirect non-trace evidence about the amount of drugs that Shonubi had transported on his seven earlier trips is that Shonubi had transported 427.4 grams on his eight' trip. The second version of Shonubi, on the other hand, is different. Here, the non-trace evidence is not individualized since the evidence which is adduced is not about Shonubi, but about other Nigerian smugglers.

6 Different ways of being naked: Shonubi and Blue Bus versus Prison Yard and Gatecrasher

Most authors present the four examples of naked statistical evidence as if they are all of the same type. In this and the next two sections, I argue that the nakedness in the Gatecrasher and Prison Yard cases differs from the nakedness of the Blue Bus and Shonubi cases. Therefore, or so I shall argue, the four cases suffer from different problems and should be analyzed in different manners.

In all four cases, trace evidence and non-trace evidence play a role in the proof of criminal facts. In the Gatecrasher and the Prison Yard case, however, the crucial piece of naked statistical evidence is trace evidence. In the Blue Bus and Shonubi cases on the other hand, the emphasis is on non-trace evidence. This difference

---

[28]Note, however, that the defense could argue that the witness is not credible because he has reasons to lie or that he is not reliable because his eyesight is bad, etc.

has implications for the analysis of the naked statistical evidence problem. Let me expound these claims in more detail.

Prison yard and Gatecrasher

In the Prison Yard case, there is quite a lot of reliable trace evidence about different aspects of the case. The camera images that are evidence of the fact that there were 100 prisoners and one guard in the prison yard at the time of the riot and that 99 of the prisoners took part in the riot and one did not, are trace evidence. The identity of the 100 prisoners is also said to be known. For example, their names might have been registered before entering the yard. Thus, the evidence that there were 100 prisoners and one guard is individualized but indirect trace evidence. More in particular, there is trace evidence about the number and the identity of all persons who had the opportunity to commit the crime. Next to that the camera images also offer direct but non-individualized trace evidence, viz that 99 prisoners took part in the riot and that one hid away. Moreover, multiple sources offer trace evidence about the crime (beating) and the (causes of) death of the guard. This too is direct trace evidence as it regards the crime, but again it is non-individualized as it does not contain information, such as fingerprints or DNA, about the individual culprits.

In the Prison Yard case it is the combination of different pieces of trace evidence that results in the naked statistical evidence that the probability that the prisoners took part in the riot is 99%. The Gatecrasher case is similar to the Prison Yard case in this respect. The number and the identity of the visitors is reliable and individualized but indirect trace evidence. The number of sold tickets is reliable direct but non-individualized trace evidence and this trace evidence results in a naked probability of more than 50%.

Shonubi

The Shonubi and the Blue bus cases are different. In the Shonubi case, there is direct and individualized trace evidence about the last trip that Shonubi made and there is also direct and individualized trace evidence that he made seven earlier trips and that he did so to transport drugs. However, the trace evidence in itself does not contain information, as regards the quantity of drugs he smuggled on the first seven trips. That is to say, the quantity of 427.4 grams is direct and individualized trace evidence in relation to his eighth trip, and thus sufficient to convict him for the crime on that trip, but it is not direct trace evidence in relation to his earlier trips. Trace evidence about the amount Shonubi's smuggled on his seven earlier trips is absent and thus there is no trace evidence that can play the same role that trace evidence plays in the Prison Yard and the Gatecrasher case.

Neither the first claim of the prosecution, that Shonubi transported the same quantity on all eight trips, nor the second claim, that he transported the same average amount of drugs as other smugglers, is based on trace evidence produced by any of Shonubi's first seven trips under consideration. The evidence that Shonubi transported 3500 grams or at least 2000 grams of heroin in total is not trace evidence in the way that 100 prisoners having an opportunity and 99 being involved in the riot is trace evidence. On the contrary, both estimates derive from non-trace evidence. In the first case it derives from individualized non-trace evidence about Shonubi's eighth trip in combination with an underlying generalization such as 'Most individual smugglers transport the same quantity on each trip they make'. Stated differently, the estimate is based on Shonubi's individual profile. In the second case, the estimate derives from non-individualized evidence, viz from evidence about what other Nigerian smugglers transported in combination with an underlying generalization such as 'Most Nigerian smugglers transport roughly the same quantity as other Nigerian smugglers'. Stated differently, this estimate is based on a group profile.

Unlike the other three cases, the dispute in the Shonubi case is not about the identity of Shonubi but about the amount of drugs he transported. This difference makes it difficult to compare the Shonubi case to the Prison Yard case. Let us therefore have a look at the Blue Bus case which illustrates the difference between the role of trace evidence and non-trace evidence more clearly.

Blue Bus

In the Blue Bus case, we have a witness statement that there was an accident and that it was caused by a bus. This is direct trace evidence about the accident. However, the claim that the accident is caused either by a bus from Blue Bus company or by a bus from Red Bus company is not based on trace evidence in the way that the Gatecrasher and Prison Yard cases are based on individualized albeit indirect trace evidence. Other than in the Prison Yard case, no mention is made of e.g. road safety cameras producing trace evidence that the accident has been caused either by a blue bus or by a red bus and not by any other bus. Betty Smith's claim that the accident can only have been caused by a red bus or a blue bus is based on non-trace evidence, viz on past experiences that only red and blue buses have been riding through town.

However, let us focus on an even more important difference. The claim that either a red bus or a blue bus has caused the accident, is in itself insufficient for the claimant to win the case, because it results in a probability of only 50%. Accordingly, the claimant needs further evidence to win her case. The choice of that evidence, the market share of the two bus companies, seems to be even more problematic than the claim that only one of the two bus companies that operate in town can be liable for the accident. It is more problematic, because it is not clear why market share is chosen as a relevant distinguishing characteristic in the first place. The market share is at best a proxy for the accident rate of the companies. One can criticize the choice of the market share for being, in the terminology of the Shonubi case, a "speculative" link between the accident and Blue Bus causing it. For example, one could argue that it would be more informative to investigate into the number of blue and red buses operating on the particular street around the time of the accident, or in the accident rate of the drivers operating these buses.

However, whichever choice of 'evidence' one would make, it could legitimately be contested and alternatives could be proposed. This choice would be much more controversial than the choice of trace evidence, such as the uncontested individualized but indirect trace evidence that 100 prisoners had the opportunity to riot and the direct but non-individualized trace evidence that 99 of them were actually involved in the riot.

In conclusion, in the Prison Yard and the Gatecrasher case it is trace evidence which produces a probability which seems high enough to convict. The trace evidence in the Blue Bus case on the other hand, i.e. the witness statement that the accident was caused by a bus, does not produce a sufficiently high probability to win the case. Therefore, the claimant appeals to non-trace evidence. This non-trace evidence, viz that the accident can only be caused by either a red bus or a blue bus and, even more dubiously, that market share is a good proxy for accident rate, is contestable. The same holds for the first line of the prosecution in the Shonubi case in which the amount Shonubi transported on his eight' trip is taken to be non-trace evidence for the amount transported on the other trips. Finally, the second line in the Shonubi case seems even more problematic since it is not based on evidence about Shonubi's own behaviour, but on non-trace evidence about the behaviour of other Nigerian smugglers.

7 Blue bus and Shonubi: non-trace evidence

In the last section I have argued that there is a difference between the Gatecrasher and Prison yard cases on the one hand and the Blue Bus and Shonubi cases on the other. I call the Gatecrasher and Prison Yard cases illustrations of a 'pure' naked statistical evidence problem. In these cases the naked statistical evidence problem results from trace evidence that on its own produces a high probability. The problem in these cases is that the trace evidence is insufficiently individualized and direct, namely only individualized but indirect (we know the identity of the 100 individuals who had the opportunity to commit the crime) or direct but non-individualized (we know that 99 of the 100 prisoners committed a crime, but we do not know their identities).

In the Blue Bus and the Shonubi cases on the other hand it is not the trace evidence (viz the accident being caused by a bus, Shonubi having made seven earlier trips) but the non-trace evidence that does most of the job of raising the probability. The Blue Bus and the Shonubi cases are not 'pure' naked statistical evidence problems. Rather, they are 'impure' because they are an entanglement of the naked statistical evidence problem and the reference class problem.[29]

8 Prison Yard and Gatecrasher: trace evidence

In section 7 I have argued that the high probabilities in the Blue Bus and Shonubi cases are mainly based on non-trace evidence and that the naked statistical evidence problem becomes entangled with the reference class problem because of the dubious choice of reference classes. In the Prison Yard and the Gatecrasher cases on the other hand, the high probability mainly derives from trace evidence and the naked statistical evidence problem does not become entangled with the reference class problem. However, this does not imply that trace evidence does not suffer from the reference class problem.

The reference class problem
Let me clarify my argument by turning the Prison yard and the Gatecrasher case into Blue Bus- and Shonubi-like cases. In presentations of the Prison Yard and the Gatecrasher case the reference class problem is ignored and it is implicitly assumed that the 100 prisoners are equal and the question is simply whether the defendants can be convicted because of the fact that 99 out of 100 (or, in the Gatecrasher, 501 out of 1000) were involved. However, by choosing a particular reference class, the probability that a particular defendant was involved in the riot (and in gate crashing) will be affected. The result might be that the probability that an individual defendant took part in the riot would fall below the required 95%.

For example, in the Prison Yard case we could follow the line of the Blue Bus case or the first line that the prosecution chose in the Shonubi case and create an individual profile for each defendant on the basis of data about the individual defendants. In the Shonubi case, the prosecution combined the individualized non-trace evidence that Shonubi smuggled 427.4 grams of heroin on his eight' trip with the individualized indirect trace evidence that he made seven earlier smuggling trips

---

[29]See [5] p. 171, note 11, p. 174-5. The authors argue that the reference class problem is not only a problem for frequentist interpretations of probability but also for other interpretations. Even if probabilities are not identified with frequencies but for example with beliefs, it can be argued that these beliefs should be 'answerable to' frequencies.

to argue that he had smuggled 8x427.4 grams in total. Similarly, in the Prison Yard case, the prosecution could create an individual profile for each defendant, using only data about him. For example, in the case against D93, the prosecution could collect data about him, e.g. whether or not he had attacked a guard or taken part in riots on earlier occasions and, combining it with an 'underlying' generalization, use it as evidence against him. This evidence would be individualized non-trace evidence.

However, we could also follow the second line that the prosecution chose in the Shonubi case, namely to use a group profile based on data about other smugglers who are unrelated to Shonubi's crimes. The prosecution used data about other drug smugglers who were caught in the same period to calculate the average weight they smuggled and used that as evidence that Shonubi had smuggled the same amount. If we take this route in the Prison yard case, we could investigate characteristics of prisoners who took part in riots in this particular prison in the past few years to create a group profile and use it to predict which prisoners are most likely to have rioted. In this case, the non-trace evidence would not be individualized. In both ways, i.e. by creating an individual profile or a group profile the prior probability that an individual defendant has taken part in the riot could rise or fall, depending on the evidence that is used.

Naked statistical evidence problem

Leaving apart the reference class problem, let us now investigate how we can analyse the naked trace evidence in the Prison Yard (and the Gatecrasher) case. We have seen that the scenario approach instructs fact finders to compare different scenarios to determine which scenario offers the best causal explanation of the trace evidence in a case. In the Prison Yard case, we must compare one or more prosecution scenarios to one or more defense scenarios. Both Cheng and Di Bello have offered a Bayesian analysis of the naked statistical evidence problem that is similar to the argument I offered in section 5.[30] Both point to the fact that naked statistical evidence cases are problematic because if one compares the prosecution scenario that the defendant took part in the riot to the defense scenario according to which the defendant hid away, the likelihood ratio of the evidence that 99 were involved and 1 hid away is 1.[31]

In section 5 I briefly compared the prosecution scenario that defendant D93 took part in the riot to the defense scenario that D93 was the person hiding away. The prosecution scenario can causally explain the camera images according to which D93

---

[30] [3] and [10].

[31] [10] argues that the likelihood ratio is not exactly 1, but for the purpose of this article we can ignore the subtleties of his argument.

is one of the 99 prisoners who took part in the riot, but the defense scenario can also causally explain the camera images and it seems it can do so equally well. So, even though the naked trace evidence that D93 took part in the riot is 99%, it does not discriminate between the prosecution and the defense scenario. Compare this to a variant of the Prison Yard case in which we have direct individualized trace evidence, for example a reliable witness who has identified D93 as one of the rioting prisoners. In principle, this piece of evidence does discriminate; depending on the reliability of the witness the prosecution scenario can offer a better or even a much better explanation of the witness statement than the defense scenario.

It is important to repeat a remark I made in section 5, viz that the question whether evidence is individualized is relative to the specific scenarios being compared. The non-individualized trace evidence in the Prison Yard case can discriminate between the scenario of the prosecution and many other scenarios, for example the scenario according to which a person who did not belong to the group of 100 prisoners took part in the riot. In relation to other persons the trace evidence is individualized since the camera images offer evidence that the perpetrators were prisoners. However, this is not relevant for the analysis of the case, since the scenario that an outsider was involved in the riot is not one of the scenarios under discussion. What matters is that the naked evidence does not discriminate between the prosecution and defense scenarios.

What do my brief remarks add to the extensive analyses of Cheng and Di Bello? Cheng and Di Bello do not discuss the differences between the Blue Bus and Shonubi cases on the one hand and the Prison Yard and Gatecrasher cases on the other. Accordingly, they do not note that their analyses only apply to cases of naked statistical trace evidence and that cases of naked statistical non-trace evidence demand a different analysis.[32] However, if we are correct to claim that naked trace evidence does not discriminate between scenarios, then it seems that a conviction in Prison Yard-like cases can only be based on non-trace evidence. Obviously, this might take us back to the three questions posed at the end of the last section, viz the reference class problem, the question to what extent it is legally acceptable that a high probability derives mainly from individualized non-trace evidence (individual profile) and to which extent there is a role for non-individualized non-trace evidence (group profile). However, it was not the aim of my paper to solve the naked statistical evidence problem, but only to argue that there are different ways of being naked. Therefore, I will leave it to others to answer these questions.

---

[32][11] offer a critical analysis of profiling evidence. On p 23 note 57 they correctly state that their arguments against the use of profiling evidence do not apply against cases like the Gatecrasher. However, they do not seem to note that this is due to the fact that Gatecrasher-like cases rely on trace evidence whereas cases like Blue Bus rely on non-trace, i.e. profiling, evidence.

9 Conclusions

In this article I have offered an analysis of the nature of naked statistical evidence. I have argued that there are at least two different ways of being naked. In the Blue Bus and Shonubi cases the high probability mainly derives from individualized or non-individualized non-trace evidence in combination with contestable choices of reference classes. In the Prison Yard and the Gatecrasher case the high probability mainly derives from a combination of individualized but indirect and direct but non-individualized trace evidence. The main problem in these cases is not the choice of reference classes, but the fact that the trace evidence does not discriminate between the relevant competing scenarios about the case.

Since my analysis relies on a comparison of four specific examples, I do not claim that my analysis covers all possible types of naked statistical evidence. My comparison of the four cases does suggest, however, that arguments for and against the use of naked statistical evidence should take the distinction between trace and non-trace evidence into account and that they should separate the naked evidence problem more clearly from the reference class problem.

References

[1] Ron J. Allen and Michael S. Pardo, Relative Plausibility and its Critics, *The International Journal of Evidence and Proof*, volume: 23 issue: 1-2, 5-59 (DOI: 10.1177/1365712718813781), 2019

[2] Lara Buchak, Belief, credence, and norms, *Philosophical Studies: An International Journal for Philosophy in the Analytic Tradition*, Vol. 169, No. 2, pp. 285–311, 2014

[3] Edward K. Cheng, Reconceptualizing the Burden of Proof, *The Yale Law Journal* 122 5, 1254–1279, 2013.

[4] L. Jonathan Cohen, *The Probable and the Provable*, Oxford, Oxford University Press, 1977.

[5] Mark Colyvan, Helen Regan and Scott Ferson, Is it a Crime to belong to a Reference Class?, *The Journal of Political Philosophy* 9 2, 168-181, 2001.

[6] Mark Colyvan and Helen Regan, Legal Decisions and the Reference Class Problem, *E&P* 11, 274–285, 2007.

[7] Christian Dahlman, Unacceptable generalizations in Arguments on Legal Evidence, *Argumentation*, 31, 83–99, 2017

[8] Christian Dahlman, Naked Statistical Evidence and incentives for lawful conduct, *The International Journal of Evidence and Proof*, Vol. 24(2) 162–179, 2020.

[9] Marcello Di Bello, *Statistics and Probability in Criminal Trials*, PhD thesis Stanford University, 2013.

[10] Marcello Di Bello, Trial by Statistics: Is a high probability of Guilt enough to convict?, *Mind*, Volume 128, Issue 512, 1045–1084, `https://doi.org/10.1093/mind/fzy026`, 2019.

[11] Marcello Di Bello and Collin O'Neil, Profile Evidence, Fairness and the Risk of Mistaken Convictions, *Ethics* (130), no. 2, 147-178 `https://doi.org/10.1086/705764`, 2020.

[12] Igor Douven, "Abduction", *The Stanford Encyclopedia of Philosophy* (Summer 2017 Edition), Edward N. Zalta (ed.), `https://plato.stanford.edu/archives/sum2017/entries/abduction/2017`.

[13] David Enoch and Talia Fisher, Sense and Sensitivity: Epistemic and Instrumental Approaches to Statistical Evidence, *Stanford Law Review* 67, 557-611, 2015.

[14] Andrew Guthrie Ferguson, Big data and predictive reasonable suspicion, *University of Pennsylvania Law Review*, 163, 2, pp. 327-410, 2015.

[15] Gilbert H. Harman, The inference to the best explanation. *Philosophical Review* 74, 88-95, 1965.

[16] A.J. Izenman, Introduction to Two Views on the Shonubi Case. in J.L. Gastwirth (ed.) *Statistical Science in the Courtroom*, Springer, 393-403, 2000.

[17] Hylke Jellema, Case comment: responding to the implausible, incredible and highly improbable stories defendants tell: a Bayesian interpretation of the Venray murder ruling, *Law, Probability and Risk*, Volume 18, Issue 2-3, 201–211, `https://doi.org/10.1093/lpr/mgz011`, 2019

[18] Liat Levanon, Statistical Evidence, Assertions and Responsibility, *The Modern Law Review*, 82(2), 269–292, 2019.

[19] Peter Lipton, *Inference to the best explanation (2nd ed.)* London: Routledge, 2004.

[20] Anne Ruth Mackor, Novel facts: The relevance of predictions in criminal law. *Strafblad*, 15, 145–156, 2017.

[21] Dana K. Nelkin, The Lottery Paradox, Knowledge, and Rationality, *The Philosophical Review*, Vol. 109, No. 3, pp. 373-409, 2000.

[22] Charles R. Nesson, Reasonable Doubt and Permissive Inferences: The Value of Complexity, *Harvard Law Review*, 92, 6, 1187-1225, 1979.

[23] Paul Nieuwbeerta and Gerlof Leistra, Dodelijk geweld: *Moord en doodslag in Nederland. [Deathly Violence: Murder and manslaughter in the Netherlands]*, Amsterdam, Balans, 2007.

[24] Nancy Pennington and Reid Hastie, The story model for juror decision making. In R. Hastie (ed.), *Inside the jury: The psychology of juror decision making* Cambridge: Cambridge University Press, (2nd ed.), pp. 192-221, 1993.

[25] Mike Redmayne, Exploring the Proof Paradoxes, *Legal Theory* 14, 281-209, 2008.

[26] Frederick F. Schauer, *Profiles, Probabilities and Stereotypes*, Cambridge Mass., Harvard University Press, 2003.

[27] Judith Jarvis Thomson, Liability and Individualized Evidence, *Law and Contemporary Problems*, 9 3, 199-219, 1986.

[28] Laurence H. Tribe, Trial by Mathematics: Precision and Ritual in the Legal Process,

Harvard Law Review, 84, 1329-1393, 1971.

[29] H. Richard Uviller, Evidence of Character to Prove Conduct, *University of Pennsylvania Law Review*, 130 4, 845-891, 1982.

[30] Peter J. Van Koppen, *Overtuigend bewijs: Indammen van rechterlijke dwalingen [Convincing evidence: Reducing the number of miscarriages of justice]*. Amsterdam: Nieuw Amsterdam, 2011.

[31] van Koppen, P. and Mackor, A. R.: A Scenario Approach to the Simonshaven Case. *Topics in Cognitive Science*, 12(4), 1132-1151. https://doi.org/10.1111/tops.12429, 2020

[32] Willem A. Wagenaar, Peter J. van Koppen, and Hans F.M. Crombag, *Anchored narratives: The psychology of criminal evidence*. London: Harvester Wheatsheaf, 1993.

 Received 2 August 2019

USING DISTANCE IN ARGUMENT MAPS TO MODEL CONDITIONAL PROBATIVE RELEVANCE

DOUGLAS WALTON
University of Windsor

Abstract

Using argument maps drawn in the style of the Carneades Argumentation System this paper analyzes a series of typical examples of legal argumentation where the relevance of evidence is an issue. It is shown how probative relevance in legal argumentation of the kind defined in the Federal Rules of Evidence can be modeled by bringing in the notion of distance from linguistics. By these means it is explained how distance in relevance is related to conditional relevance.

1 Introduction

The aim of this paper is to take some first steps toward modeling the notion of legal relevance defined in the Federal Rules of Evidence (FRE), by using argumentation schemes and argument maps drawn as graphs in the style of the Carneades Argumentation System . The graphs are applied to examples in order to link distance to conditional relevance as defined in the FRE. This approach, I hope, might be of interest to legal professionals or scholars in other fields (such as linguistics and philosophy) where relevance is a needed but worryingly vague concept. The project aims to clarify relevance by attempting to make it more precise as a first step towards developing it into a useful concept for recent work in artificial intelligence. [13] (p. 1) uses three factors to assess the relevance of an utterance in argumentation: (1) the number of inferential steps required, (2) the types of argumentation schemes involved, and (3) the implicit premises required. These three factors are used to define inferential distance [13] (p. 16): "Inferential distance can be defined as the number (quantity) and acceptability (quality) of the argumentative inferences needed for connecting a premise to a conclusion". More specifically, "relevance can be more or less acceptable or effective from an argumentative point of view" [13] (p. 16), depending on "the probative force of the type of argument and the number and acceptability of the needed implicit premises or intermediate conclusions".

This notion of distance becomes very interesting with respect to modeling relevance in law if you take the view [25] that determining the relevance of one argument to another in legal argumentation typically depends on an intervening sequence of argumentation from some evidence to an ultimate conclusion at issue. In this paper, we will be concerned with the relevance of arguments to a proposition or vice versa. The focus is on examples of legal argumentation where there is supposed to be a sequence of argumentation of the kind that can be visually represented as an argument map, a directed graph, that moves towards proving, or giving evidence to support a particular proposition that is the so-called ultimate *probandum* in a case at trial. The question is how is one is supposed to track through the graph from the given evidence to the ultimate proposition to be proved in a trial. This paper works out an answer by using Macagno's notion of inferential distance as modeled by Carneades.

Section 2 briefly outlines the historical background where earlier writers attempted to draw a distinction between logical relevance and legal relevance. Especially significant are the views of J. H. Wigmore on the subject because he actually used an argument mapping method to represent [1] evidential reasoning and relevance in real legal cases. Section 3 draws a distinction that is a necessary preliminary step to any attempt to investigate relevance in argumentation. This is the important distinction between probative relevance and topical relevance. Section 4 states and briefly comments on the rules in the Federal Rules of Evidence (FRE) that provide criteria for judging whether evidence is relevant in legal cases. Section 5 shows how Carneades uses standard argumentation tools such as argumentation schemes, critical questions matching schemes, argument maps, and chaining of arguments, tools that can potentially be used to model relevance of argumentation. Section 6 outlines some formal models of legal argumentation that employ or depend on the concept of relevance of argumentation. Section 7 shows how relevance needs to be defined in the context of a trial. A burden of proof, the so-called burden of persuasion in the common law, is set in place at the opening stage, and that is used to decide whether an argument is relevant or not by weighing the arguments on both sides. Section 8 introduces the notion of conditional relevance as defined in the FRE. Section 9 discusses the modeling of probative weight of the kind used in the Federal Rules of Evidence. Section 10 provides some conclusions.

---

[1]http://carneades.github.com/

2   The History of Legal Relevance

Part of the problem of legal relevance is to draw a distinction between logical relevance of a general sort that can apply to different kinds of settings in which argumentation takes place and legal relevance, which should be specifically defined to model something that is widely employed in law as one of its most fundamental tools. Textbooks on evidence law have acknowledged this distinction. For example, [17] (p. 125), characterized logical relevance very broadly as referring to evidence "that has any tendency in logic to establish a proposition". The history of evidence law supports the view that legal relevance should be based on logical relevance. Wigmore's treatise, Evidence in Trials at Common Law ([28],vol. 1A, p.1004-1095) in the footnotes written by Peter Tillers, summarizes how Wigmore's views on relevance developed from the writings of earlier evidence scholars.

[16] made a fundamental point when they postulated a distinction between direct and indirect relevance. In their view, an evidential proposition can be directly relevant to a conclusion if it is probative of it in a single step of proof, or indirectly relevant to it by a "transitive inferential relation from one step to another" ([16], Part I, 1279). [16] (Part I, p. 1279) saw relevance as "entirely a matter of logic". Wigmore only partly agreed. According to Wigmore there is a science of proof, meaning an underlying structure of logical reasoning, but he went on to distinguish between relevance as a logical notion and judgments of relevance of the kind made in a trial. Wigmore held that the rules of relevance that apply in a judicial tribunal, while they are unique to this setting, are based on a logical notion of relevance underlying legal reasoning [20] (p. 156).

What is especially remarkable about Wigmore's theory of relevance is that it was illustrated in specific cases by his use of so-called "evidence charts" which were essentially argument maps of the kind currently being used in computational argumentation and informal logic. Wigmore's theory of evidence was built around the central notion of a chaining of inferences visualizing the mass of evidence as a set of propositions in a case leading to one proposition designated as the ultimate *probandum*. He represented the concept of relevant evidence using argument maps [27] that had the following properties. The ultimate *probandum* in the trial is represented by a proposition which appears in an argument map at the root of the tree. The remaining nodes in the tree represent propositions that are items of evidence supporting or attacking the ultimate proposition, such as instances of witness testimony or circumstantial evidence. Legal relevance as, Wigmore viewed it, can be defined with reference to this kind of abstract graph structure. A proposition or argument should count as relevant evidence in a trial if it fits in as a node in the tree leading to the ultimate *probandum*. [2] This view presents the outlines of the Wigmorean theory of legal relevance, a theory that is highly compatible with current computational models of argumentation, such as ASPIC+ and Carneades. These models too use argument maps.

3 Probative Relevance v. Topical Relevance

One propositions is *probatively relevant* to another if one can be used to support or attack the other as part of an argument (or sequence of arguments) that supports the other or casts reasonable doubt on it. Probative relevance has three properties as a relation. It is reflexive (arguably), non-symmetrical and transitive.

Reflexivity: proposition P is always probatively relevant to itself.

Non-symmetry: if P is probatively relevant to another proposition Q, it does not necessarily follow that Q is probatively relevant to P.

Transitivity: if P is probatively relevant to Q and Q is probatively relevant to R, then P is probatively relevant to R.

Also note that (1) an argument can be probatively relevant to a proposition (such as a proposition at issue) and (2) one argument can be relevant to another. How this works will be illustrated in some examples below.

This is a good time to draw an important distinction between direct and indirect relevance. One argument may be said to be directly relevant to another in an argument if the first one presents evidence that probatively supports or attacks the other in one step. Essentially this means that there is an argumentation scheme into which the premises and the conclusion of this argument fit. But one argument can be indirectly relevant to another if the two are connected by a sequence of arguments with more than one step so that one can be used to support or attack the other through a sequence of intervening arguments, as shown by the Carneades examples in the figures in this paper.

In law, propositions are probatively relevant if one can be used to support or attack the other, using a sequence of argumentation going over a distance, a sequence of steps, from the one to the other, and based on some evidence that has already been accepted as factual and admissible in a discussion or procedure. Such a sequence may require only one inferential step, or it may require many.

Probative relevance needs to be contrasted with topical relevance. The two propositions âĂŸBananas are yellow' and âĂŸOver 20,000 bananas were imported from Honduras to Ontario in 2015' are topically relevant to each other because both are

---

[2]Wigmore's theory of how argumentation moves forward in a trial through five stages is explained in section 7

about bananas. They have subject- matter overlap and so one proposition is topically relevant to the other [26] (p. 554). However, these two propositions are not probatively relevant to each other, because you can't prove or disprove the one from the other. It may be true that over 20,000 bananas were imported from Honduras to Ontario in 2015, the truth of that proposition does not support or refute the proposition that bananas are yellow [3] Of course one could be probatively relevant to the other if it could be found that there was a sequence of arguments connecting the two and each of the arguments fitted an argumentation scheme, or could otherwise be shown to be valid. But as far as we know from the textual evidence in this example, there was no such connection.

In contrast with probative relevance, topical relevance has the following three properties.

Reflexivity: proposition P is always topically relevant to itself.

Symmetry: if P is topically relevant to Q, it follows necessarily that Q is probatively relevant to P.

Failure of Transitivity: if P is probatively relevant to Q and Q is probatively relevant to R, then P may or may not be probatively relevant to R.

The first two properties are relatively obvious, but even though transitivity has been more controversial, [5] (p. 77) offers the following example to show that transitivity does not hold. The proposition that John loves Mary is relevant by topical overlap to the proposition that Mary has two apples, and the second proposition is relevant to the third proposition that 2+2 = 4. But if topical relevance were to have the property of transitivity, âĂŸJohn loves Mary' would be relevant to âĂŸ2+2 = 4'. [5] (p. 77) expresses doubt about this possibility as follows: "It may be comforting to mathematicians to think that love and mathematics have something in common, but for most of us that sounds like sophistry".

[12] have shown how changes of topic in a piece of natural language text can be used in argument mining as evidence to help with determining whether statements are relevant to an ultimate *probandum* in a tree structure. It is assumed by this method that the argumentation can be represented as a tree where the conclusion is given first and then a line of evidential reasoning is followed supporting this conclusion [12] (p. 129). Relevance is determined by examining how similar the topic of each proposition is to its predecessor. If the similarity runs along the line of reasoning being followed, this finding can be used as an indication of relevance. So although topical relevance and probative relevance always need to be clearly distinguished, there can be connections between the two in real argumentation, and these connections can be important for computational modeling of relevance.

---

[3]Richard Epstein ([4]) built a formal system of relatedness logic based on this idea.

4 Relevance and the Federal Rules of Evidence

Some indication of how the Federal Rules of Evidence [4] are applied to judge relevance in the setting of a legal trial can be given by briefly summarizing the account given in [19] (chapter 16). To pursue the goal of finding the truth underlying the dispute in a trial, the law of evidence strives to present all the evidence that bears on the issue to be decided, or at least all the evidence that is judged to be relevant. The presumption is that unless there is a ground for refusing to hear such evidence, it should be considered admissible [19] (p. 773). In general however, relevance determines admissibility of evidence. This means, in logical terms, that evidence is admissible if and only if it is relevant.

In law, there are two components of relevant evidence, materiality and probative value. Materiality concerns "the relation between the propositions for which evidence is offered and the issues in the case" [19] (p.773). Evidence is material if and only if it helps to prove a proposition that is "a matter in issue". What is an issue is determined by the pleadings. For example, "in a suit for worker's compensation, evidence of contributory negligence is considered to be immaterial since the worker's negligence does not affect the right to compensation" [19] (p. 79). This remark indicates that relevance is affected by the elements of the issue.

How is relevance tested for in a given case at issue? The two most important FRE rules for this paper are Rules 401 and 403.

> Rule 401: (Test for Relevant Evidence): Evidence is relevant if (a) it has any tendency to make a fact more or less probable than it would be without the evidence, and (b) the fact is of consequence in determining the action.

> Rule 403: The court may exclude relevant evidence if its probative value is substantially outweighed by a danger of one or more of the following: unfair prejudice, confusing the issues, misleading the jury, undue delay, wasting time, or needlessly presenting cumulative evidence.

Despite its use of the word âĂŸprobable', Rule 401 is not based on any statistical or Bayesian notion of probability, at least of any sort indicated in the FRE, but on two component concepts sometimes called materiality and probative weight [19]. Sometimes probative weight is also called probative value [22] (p. 256). Materiality, to quote [19] (p. 773) once again, refers to "the relation between the propositions

---

[4]Here you can find the 2018 version of the Federal Rules of Evidence: https://www.rulesofevidence.org/

for which the evidence is offered in the issues of the case" This is good to keep in mind, because the notion of materiality as described in FRE 401 is closely related to what is called the notion of distance in this paper.

5 Examples of Probative Relevance Modeled in Carneades▮

Here is a simplified version of the scheme for argument from expert opinion adapted from [24] (p. 310). D is a subject domain of knowledge and A is a proposition.

Major premise	E is an expert in D containing A.
Minor Premise	E asserts that A is true (false).
Conclusion	A may be presumed to be true (false).

There are six basic critical questions matching the scheme for argument from expert opinion [24] (p. 310).

Expertise Question: How credible is E as an expert source?

Field Question: Is E an expert in the domain S that A is in?

Opinion Question: What did E assert that implies A? Trustworthiness Question: Is E personally reliable as a source?

Consistency Question: Is A consistent with what other experts assert? Backup Evidence Question: Is E's assertion based on evidence?

If a critic asks any one of the six critical questions, the original argument is tentatively suspended until the proponent question has replied to the question adequately. Carneades models critical questions by treating them as implicit premises of the argumentation scheme. There are two kinds of implicit premises called assumptions and exceptions, the difference between them being their burden of proof requirements. Asking a critical question classified as an assumption automatically defeats an argument. Asking a critical question classified as an exception only defeats an argument if evidence is given to back up the exception as holding. Carneades uses built-in argumentation schemes and critical questions to evaluate and produce arguments [7]. In a bipartite graph, the set of nodes are partitioned into two subsets, round nodes and rectangular nodes, in such a way that no two round nodes are adjacent to each other and no rectangular nodes are adjacent to each other.

The scheme for the argument from expert opinion is represented in the round node labeled +ex, where the plus sign indicates that it is a pro argument. Since all three premises of this argument have been accepted by the audience, they are shown in rectangles with a green background. For this reason, and because the argument fits the scheme for argument from expert opinion, the conclusion, the proposition

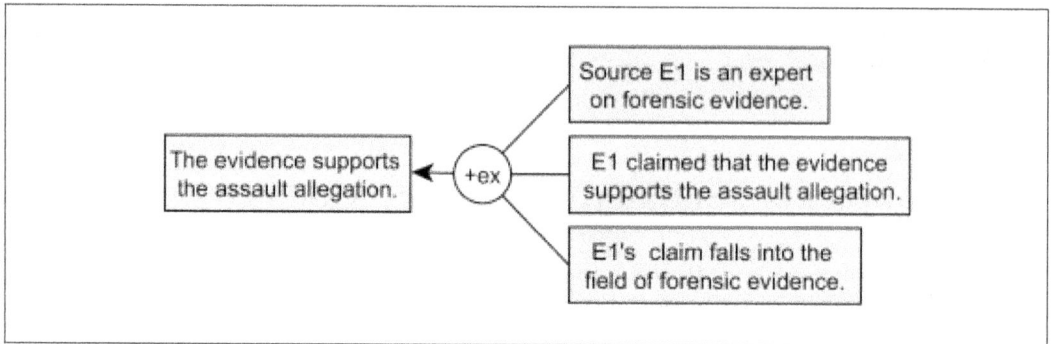

Figure 1: Carneades Map of an Argument from Expert Opinion.

stating that the evidence supports the assault allegation, is also shown in a green background. For these reasons, the argument from expert opinion shown in figure 1 is relevant to support the conclusion that the evidence supports the assault allegation by bringing in relevant evidence supporting one of the premises of the prior argument.

To prove or disprove that a specific argument or proposition is relevant can be complex because many gaps in the argumentation have to be filled in between the argument being considered and the ultimate conclusion that is supposed to be proved. Using a simple abstract example it can be indicated how a premise, conclusion or argument can be relevant in a legal case using a graph structure of version 2 of the Carneades Argumentation System [6], [25] . To illustrate this lesson we adopt the terminology that an argument is properly said to carry probative weight in support of its conclusion if and only if the argument is valid and its premises are accepted, but we do not require deductive validity because it is important in studying relevance in law to consider defeasible arguments that give probative weight towards supporting a conclusion leading to tentative acceptance of the conclusion that may have to be withdrawn later if new evidence comes in. So in Fig. 1, for example, the argument represented in the round nodes can have defeasible argumentation schemes, such as an argument from expert opinion or abductive argument.

Suppose that P (the alleged perpetrator) has been charged with the murder of V (the victim) based on evidence from three witnesses. Some bits of flesh were found under V's fingernails and DNA testing indicated that the flesh was that of P. An expert witness W1, a forensic expert, testified that DNA testing had shown that the flesh matched P's. Another expert, W2, a medical doctor, testified that V died by asphyxiation from choking. Also, a bystander, W3, who had a clear view of the exit from the building where the homicide took place, testified that he saw P leave the building just after the time of the crime.

(1) W1 testified that the flesh found under V's fingernails matches P's DNA.

(2) W1 is a forensic expert in DNA testing.

(3) The flesh found under P's fingernails matched V's DNA.

(4) If the flesh found under P's fingernails matched V's DNA, then P choked V.

(5) W2 is a medical expert.

(6) W2 testified that V died by asphyxiation from choking.

(7) W3 says he saw P leave the crime scene.

(8) W3 had a clear view of the crime scene.

(9) P was at the crime scene.

(10) P killed V.

Three schemes are used: argument from expert opinion, argument for witness testimony and defeasible modus ponens. Here is a simple version of the scheme for argument from witness testimony adapted from [24] (p. 310), represented by the notation wt in Fig. 2.

Major premise	Witness W says that proposition A is true.
Minor Premise	W is in a position to know whether A is true.
Conclusion	Therefore, A may be presumed to be true.

In Fig. 2 the propositions that are accepted by the audience (jury or judge) are shown in green.

There are three kinds of arguments represented in Fig. 2, single-premised arguments, such as a6, linked arguments and convergent arguments. A linked argument such as a1, is a multi-premise argument in which both (all) premises are taken together to support the conclusion. A convergent argument is one in which each premise represents a separate line of argument for the conclusion that can be considered as providing support for the conclusion independently of the support provided by the other premise. In version 2 of Carneades, two arguments are used to represent a convergent argument. [5] Arguments a4 and a6 are shown this way. But in some cases links in the sequence are not made explicit, and it is left up to the

---

[5]In version 4, one argument can suffice, by associating it with a weighing function which simulates convergent arguments, but these arguments are visualized in the same way as linked arguments.

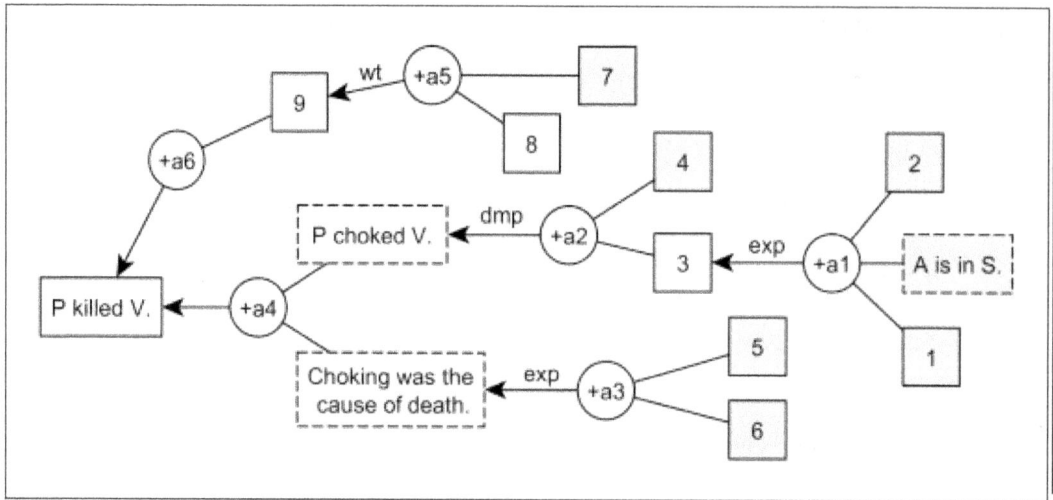

Figure 2: Relevance in the Choking Example.

audience, using their common knowledge, to fill in the gaps by specifying the intervening steps. These kinds of arguments are commonly called enthymemes, meaning that to make sense of them you have to fill in unspecified premises or conclusions in the chain of arguments that are presumed to hold. The two implicit premises in argument a4 are shown in rectangles with broken lines.

Arguments a1, a2, a3 and a4 are relevant, because in each instance there is a sequence of chained arguments leading from them to the ultimate proposition âĂŸP killed V', and each of these arguments carries probative weight that supports the ultimate conclusion. Arguments a5 and a6 are not relevant. A sequence of argumentation joining them to the ultimate conclusion exists, but premise 8 is not accepted, so the argument from a5 to a6 does not carry probative weight that supports the ultimate conclusion.

Premises 1-6 are relevant, because they are used in arguments that carry probative weight whereas premises 7, 8 and 9 are not relevant, because they do not. For an argument or a proposition to be relevant, there has to be a sequence from it to the ultimate conclusion, and the acceptance of it has to propagate along the sequence to the ultimate conclusion to support the ultimate conclusion. Here for simplicity I consider only support and not attack, but the notion of relevance can be extended to take attacking arguments into account. [6]

---

[6]In Carneades 4 a con argument is simply a pro argument for another option of some issue. Thus, in a sense all arguments are pro arguments. They are all arguments pro some option (position) of an issue. Arguments conflict if they are arguments pro different options. This is also true in

As well as showing how arguments support other arguments, Carneades, like other computational argumentation systems, also uses argument maps to show arguments can attack other arguments and defeat them. Carneades has pro arguments as well as con arguments. As shown in the examples above, Carneades indicates that a proposition is accepted by coloring its node green. It can also show that a proposition is rejected by coloring in its node red. A node with no background, a white node, indicates that the proposition is neither accepted nor rejected. A limitation of this paper is that it will only use simple argument graphs that contain green nodes to indicate acceptance of proposition and pro arguments, indicated by a plus sign in an argument node. This simple way of visually representing the distance in a sequence of argumentation from evidence to an ultimate *probandum* will make for ease of exposition.

Another example discussed in [26] (p. 11-12), which was from [14] (p. 538) concerns a trial where the lawyer for the defense objected on relevance grounds.

> Prosecutor: Mrs. Higgins, were you at the bank on June 1, one week before the robbery? Mrs. Higgins: Yes.
> Prosecutor: Did you see the defendant that day?
> Defense lawyer: Your Honor, we object on relevance grounds. May we approach?
> Judge: Yes [Lawyers come to the bench.] Prosecution, where's this testimony going?
> Prosecutor: The witness will testify that she saw the defendant outside the bank that day, that he was looking at the bank, but never came inside.
> Judge: What does that prove?
> Prosecutor: It shows that the defendant was already planning the robbery and casing the bank.
> Defense lawyer: The fact that he was outside the bank hardly proves he was planning to rob it.
> Judge: The objection is overruled. The witness may answer.

In this case the judge asked the prosecutor to make a connection between the issue of whether bank robbery was committed and the testimony of the witness elicited by the prosecutor that she (the witness) saw the defendant outside the bank.

Carneades 2. An argument con a statement P was also an argument pro not P, i.e. pro the negation of P. Conversely, an argument pro P was also an argument con the negation of P. The only difference between Carneades 2 and 4 in this regard is that in Version 2 issues had exactly two options (P, not P), while in Version 4 there can be any finite number of options (0, 1, 2, 3, or more).

Here is a simple version of the scheme for argument from practical reasoning (pr in Fig. 3) from [24] (p. 94âĂŞ95).

Major premise	I (a rational agent) have a goal G.
Minor Premise	Carrying out this action A is a means to realize G.
Conclusion	Therefore, I ought (practically speaking) to carry out this action A.

The way practical reasoning is used in this case does not conform to the requirements of the simplest scheme above. Instead, it fits into a scheme that has this form.

Major premise	Carrying out action A is a means for agent ϕ to achieve goal G.
Minor Premise	ϕ carried out action A.
Conclusion	ϕ was planning to achieve goal G.

This variant of the practical reasoning scheme can be called argument from means and action to plan. Two instances of it are shown in Fig. 3.

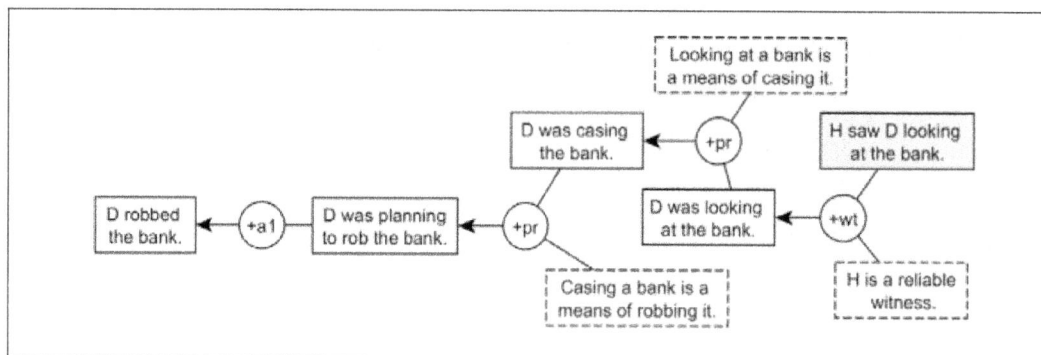

Figure 3: Bank Robbery Example

The prosecutor argued that the testimony is relevant evidence that the defendant was planning the robbery and was "casing" the bank. Based on this argument, the judge overruled the irrelevance objection. The proposition that H saw D looking at the bank is shown in a green rectangle, indicating that it has been accepted as evidential. This way of representing the sequence of argumentation in the bank robbery example shows how much of the argument from the evidence to the ultimate conclusion, shown at the left, depends on implicit premises, indicated by the rectangles with broken borders. Let's assume that the proposition that H is a reliable witness is accepted. Then it will follow, using the scheme for argument from witness testimony, that D was looking at the bank. Once this conclusion is accepted, we have to examine the premise that looking at a bank is a means of casing it. This could be accepted as a common sense generalization. If it is accepted, then the conclusion that D was casing the bank follows. But it does follow necessarily. As the defense lawyer pointed out, that D was outside the bank hardly proves he was planning to rob it.

The next inference in the sequence depends on another common knowledge generalization, the statement that casing the bank is a means of robbing it. By this means the argumentation from the evidence can be chained forward so that ultimately the conclusion would be shown to be acceptable, and so Carneades will automatically calculate forwards, showing all of the propositions, including the ultimate conclusion, as colored with a green background. This way of analyzing the structure of the argumentation is consistent with the judge's ruling that the testimonial evidence is relevant. One parenthetical remark about this way of analyzing the bank robbery example is that the proposition that D robbed the bank may not itself be the ultimate *probandum* in a real legal case. There has to be a charge, such as armed robbery for example, and D's planning to rob the bank would be evidence supporting an element of the crime of armed robbery such as intent.

How does this way of handling these two examples fit with Macagno's way of relating relevance to inferential distance? In Fig. 2, as well as Fig. 3, there is a sequence of argumentation starting from some propositions accepted as evidence and moving forward to an ultimate conclusion that is at issue in a trial. The subarguments fit argumentation schemes. Implicit premises have to be added in order for the argumentation to make sense. These components all seem to fit with Macagno's use of the notion of inferential distance to evaluate arguments for relevance or irrelevance. Could we represent inferential distance by the number of inferential steps it takes to get from the given argument to the ultimate conclusion? So, for example, the inferential distance between the proposition that H saw D looking at the bank to the proposition that D robbed the bank can be counted as the number of arguments required to get from the one to the other? In this instance the distance would be four inferential steps. But there is more to it than this. Distance also depends on whether the intervening premises are "unacceptable or poorly acceptable" and whether the "inferences that proceed from premises or result in intermediate conclusions can be doubtful" [13] (p. 16). Note in this case that defense lawyer argued that the fact that D was outside the bank hardly proves he was planning to rob it. This statement clearly shows a weakness in the argument, but one that was rebutted by the prosecutor.

Next we need to take a broader view by looking at how relevance might be formally modeled in different ways, and by seeing how how there are special features of a legal trial that indicate how relevance needs to be defined in this setting in a pragmatic way.

6 Formal Argumentation Systems for that Model Relevance

One of the most important features of relevance in the theory of relevance of [22] is its contextual variability, meaning that the relevance or irrelevance of an argument depends on the type of dialogue that the arguers are supposed to be engaged in. The six basic types of dialogue formulated in [23] are: persuasion dialogue, deliberation, inquiry, negotiation, information-seeking and eristic dialogue (quarreling). So, for example, the making of a threat might be acceptable as relevant in a negotiation dialogue, such as in union management negotiations and a strike situation, but the same ad baculum argument based on the making of a threat might rightly be judged to be irrelevant in a persuasion dialogue. The branch of linguistics concerned with the context of use of language is called pragmatics. Hence the [22] Walton theory of relevance is pragmatic in this sense, meaning that the context of use of an argument in a particular type of dialogue setting is an important part of the evidence used to judge whether an argument is relevant or not in a given case.

In a given natural language text containing argumentation it can be not straightforward in some cases to see whether a particular argument should be classified as occurring in a persuasion dialogue or a deliberation. There can be instances of persuasion over action, cases where the proponent is trying to[1] (115). However, persuasion dialogue has the ultimate goal of resolving a conflict of opinions, whereas deliberation has the goal of reaching a decision on what do in circumstances requiring a choice. Persuasion dialogues are much more adversarial than deliberation dialogue [11] (p. 33). In a deliberation dialogue, the participants cooperate to try to find a decision that is best for the group, even if it means sacrificing their individual interests [?]. In a persuasion dialogue, any argument put forward has a burden of proof attached to it, but in a deliberation dialogue, there is only a burden of responding appropriately to a proposal by explaining the reasoning behind it.

[18] has presented a formal argumentation system that models relevance. In the [23] model, replies may not be postponed. In the Prakken system, each of the two parties may make multiple moves at each turn, and it is possible to postpone replies ([18], p. 290). A benefit of this approach is that the parties can select and choose what arguments to develop in the replies, enabling them to focus on defending or attacking arguments that they think could be decisive. This might lead to a greater depth of argumentation by relaxing the greater rigidity imposed by the turn-taking rules of the Walton and Krabbe model of persuasion dialogue, even though [23] have generally two kinds of persuasion dialogues, permissive persuasion dialogues (PPD dialogues) and restrictive persuasion dialogues (RPD dialogues).

A move is relevant in a Prakken persuasion dialogue if it would constitute a winning of the dialogue for its proponent if the respondent's move that it targets is immediately defeated. Note that this type of Prakken persuasion dialogue is based on abstract argumentation [3]. This means that relevance of any given move can be determined by the purely semantic criterion of whether the move that it targets is immediately defeated or not. [18] (p. 296) explicitly states that dialogues of this sort have no context: "the framework abstracts from the communication language except for an explicit reply structure". From a logical point of view this property might seem to be a decisive advantage, because it can be immediately calculated in the abstract argumentation model whether any given argument is relevant or not, even before the dialogue has been completed. On the other hand, the abstract argumentation model only evaluates relevance of arguments, and has to be extended in order to model the relevance of propositions, such as premises or conclusions in an argument, or other speech acts in a dialogue.

Prakken's idea of relevance seems conceptually different from the idea of relevance of (Walton, 2004) and [26]. Prakken's idea is that an argument is relevant if it immediately defeats or reinstates another argument that puts the other side in a winning position. But on the approach of Walton and Macagno there is a distinction between direct and indirect relevance, where direct relevance happens immediately after the manner of Prakken, but where there are also cases involving a lengthy sequence of argumentation between the two arguments that are supposed to be relevant to each other or not. It remains to be seen how significant these differences are. But it should be kept in mind that ASPIC+ has a different conception of argument from Carneades. In ASPIC+ an argument is a tree, a subset of the argument graph, not just a single argument node. Another formal dialogue system in which relevance can be defined [11] combines the MHP model of deliberation with the Prakken model of persuasion dialogue. Deliberation is an important type of dialogue in law, but in this paper we will direct our primary attention to relevance of argumentation in trial setting in the common law system. In this setting a burden of proof called a burden of persuasion is set in place at the opening stage. Once this is set in place, it is the main factor that determines what is relevant or not as the argumentation stage of the trial proceeds. But this is a general point that needs more elaboration. Once an argument has been concluded, it is comparatively easy, in hindsight, to tell which arguments were relevant and which were not. But the basic problem is that in the middle of a discussion, it may not be at all easy to see or predict where a line of argument may be leading. It may sometimes be difficult to judge whether the argument will turn out to be relevant or not. It is common for a judge to ask the two lawyers to approach the bench and ask the one who used the argument in question where this line of reasoning is going. The lawyer may reply that if the judge will give him a few minutes they will be able to see how the argument is relevant.

7   Probative Processes in a Trial

According to the ancient *stasis* (also called status in Latin) theory [9] a speech designed to persuade an audience aims at proving or disproving a single claim called the ultimate *probandum*. This proposition represents one side (pro or con) concerning the issue to be debated in the speech. Once this proposition is set in place, prior to the beginning of the debate, it determines which arguments in the speech are relevant and which can be dismissed as irrelevant. The standard kind of example used to illustrate how stasis theory works is that of a criminal trial ([10], [9]). When a defendant is charged with theft, the ultimate issue is whether he committed the crime of theft or not.

In a trial, in order to decide the ultimate issue by weighing the arguments on both sides, a burden of proof, the so-called burden of persuasion in the common law, is set in place at the opening stage of the trial. Relevant arguments are ones that support or attack the elements of the crime of theft. One of these is the proposition that what was taken was some property that did not belong to the defendant. Another is the proposition that the defendant intended to permanently deprive the owner of this property. Once this ultimate issue to be decided is set in place by the pleadings prior to the trial, it stays in place right through the whole trial to the closing stage. The only way to change it is to dismiss the current trial and move to a new one. This property is preserved in version 2 of Carneades, as this system applies to argumentation in a trial.

The overall structure of the argumentation in a trial follows the general pattern shown in Fig. 4. At the far left the ultimate *probandum* (Prob), such as the proposition that the defendant committed theft, is indicated. Let's say for example that there are three elements. Each of these elements is necessary to be proved in order to prove Prob, and when all of them are proved, that is sufficient to prove Prob. The standard of proof to be applied, such as that of beyond reasonable doubt or that of the preponderance of the evidence, applies to Prob and individually to each of the three elements EL1, EL2 and EL3. (EL1 & EL2 & EL3) is a sufficient condition for rational acceptance of P, plus each one is a necessary condition for rational acceptance of P. This shows that the relation of the elements to the *probandum* is not merely one of conjunction.

Typically, as well, there are arguments given to support EL1, EL2 and EL3. In Fig. 4 these three arguments are a2, a3 and a4. Each of these arguments has premises, and some of these premises are evidential propositions, propositions that have been

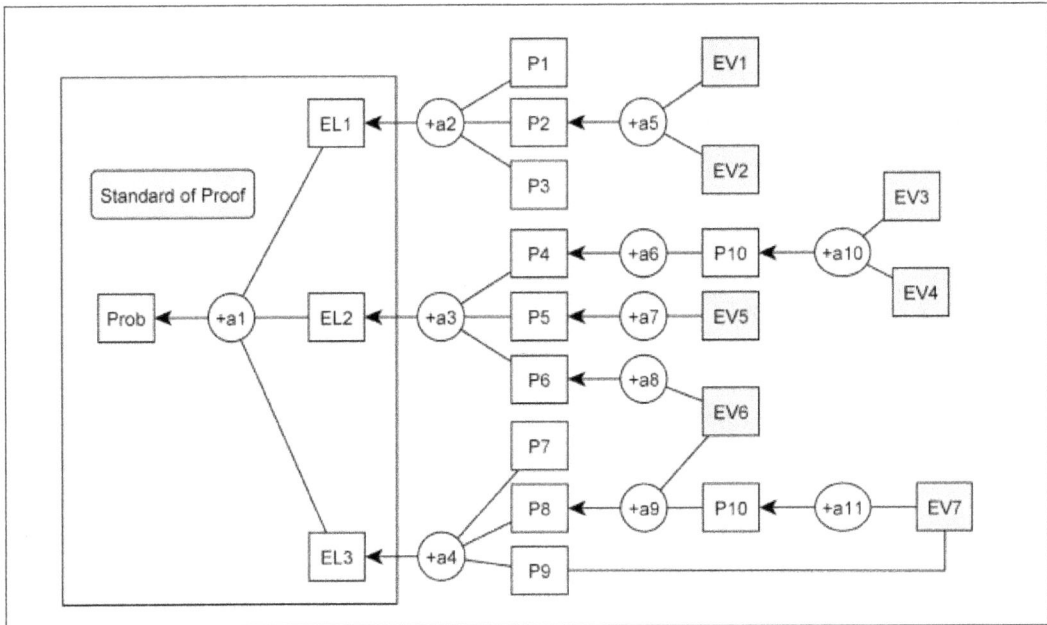

Figure 4: Standards of Proof and Elements in a Trial

previously accepted by the trier of fact as evidence. These evidential propositions are shown in rectangular nodes that have a green background. Their relevance is shown by how they propagate forward to one of the elements [8]. For example the argument a3 is relevant to EL2 because it can be proved using the sequence of argumentation that supports it, whereas the argument a4 is not relevant because there is no evidence given to support either premise P7 or P9. However, the argument a4 could be said to be conditionally relevant (see section 8) if P7 and P9 were to be supported by evidence sufficient to merit their acceptance.

The part of the argumentation in a trial shown in Fig. 4 is only one part of a sequence of argumentation that goes through five stages, according to the account given in Wigmore's *Principles* (1931) and summarized by [20] (p. 129). In this account the first stage is the part of the argumentation where the proponent asserts facts in order to provide evidence for an ultimate *probandum*. In the second stage, the opponent offers an explanation of other facts that take away the value of the arguments put forward in the first stage. In the third stage the opponent denies the evidentiary facts on which the proponent's arguments are based. In the fourth stage the opponent's arrival facts are adduced against the ultimate *probandum*. In the fifth stage the proponent corroborates facts that negate the opponent's explanation. [20] (p. 129) offers a simple example, adapted from Wigmore's textbook for students.

To charge the defendant with homicide, the prosecution offers an old quarrel, a recent thread and blood traces on clothes. Next, the defendant explains away the old quarrel by giving evidence of a reconciliation, by giving an alternative explanation of the blood traces, by showing a recent killing of a chicken, and by denying the fact of the threat. At the next stage the defendant presents rival facts of an alibi and a character for peacefulness. This account of the different stages the argumentation in a trial goes through is of interest to note here because it shows that only the first stage is represented by the argument map shown in Fig. 4. To represent the other four stages, not only do con arguments have to be represented as well as pro arguments, but it has to be understood how each stage has its own special features, and how the stages are woven together coherently into a dialogue exchange that is ordered by rules concerning who can say what and when. In addition, the notion of explanation has been brought in, raising the general issue of whether an explanation can be relevant to an argument and vice versa. These considerations clearly take us well beyond the kind of framework represented in Fig. 4. These issues have been addressed in the recent work of [2] in his theory of script-based evidential reasoning in law which extends the evidential mapping techniques used in this paper by joining explanations to arguments.

8 Distance and Probative Relevance

Transitivity of probative relevance, as shown in section 3, is related to another defining property of probative relevance. The transitivity property refers to what is called chaining of arguments in computing and the serial pattern of arguments in informal logic. This principle can be generally expressed as follows in logic. Let's call it the rule of transitive closure of probative relevance: if p1 is probatively relevant to p2, and p2 is probatively relevant to p_m, ..., and p_m is probatively relevant to p_n, then p_1 is probatively relevant to p_n. Let's identify p_n with the ultimate proposition to be proved (or ultimate *probandum*), as defined by the issue in a given case. A sequence of argumentation illustrating distance is shown in Fig. 5, where P12 is the ultimate *probandum*. The two propositions at the right are accepted (green, shaded). Let's assume that the standard of proof for P12 is high enough to require proof of both a9 and a10.

As represented in Fig. 5, a9 and a10 are two convergent arguments supporting P12 while a7 is a linked argument and so is a8. In Fig. 5, the two propositions P1 and P2 have been accepted. These are designated as evidential propositions. We also assume for purposes of illustration that all the arguments shown in the figure are valid. The problem is to figure out whether P1 is probatively relevant and whether

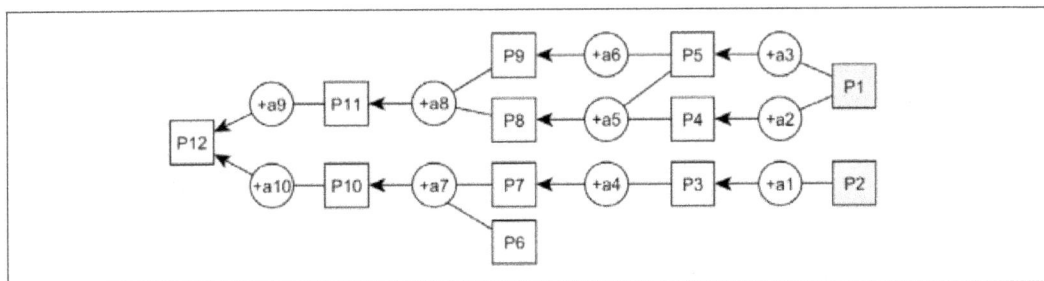

Figure 5: Evidential Propositions in a Carneades Argument Map

P2 is probatively relevant. Let's consider P1 first.

Look at the top sequence of argumentation starting at P1 and moving left. Since

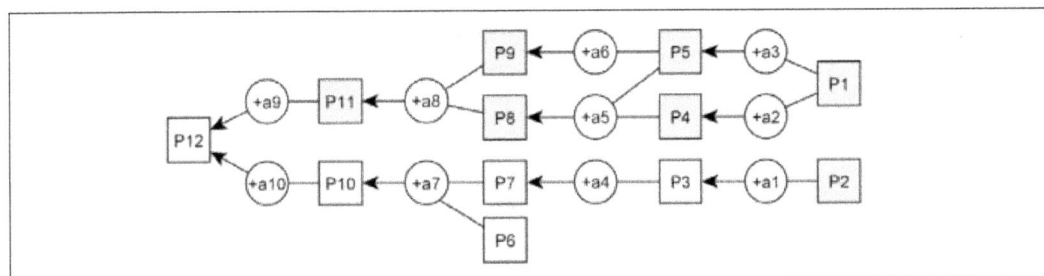

Figure 6: Propagation of Evidence in a Carneades Argument Map

both P4 and P5 are accepted, P8 is accepted. Since P5 is accepted, P9 is also accepted. Therefore P11 is accepted. But P12 is not accepted. The reason is that the standard of proof for P12 is too high for a9 to prove it. So P1 is not relevant, because its relevance depends on another factual proposition that has not been proved yet, namely P10. P12 will be colored green only if all its premises are accepted and the argument is valid. This means that as things stand according to the evidential picture given in Fig. 6, P1 is not probatively relevant with respect to the *probandum* P12.

However, P1 could be said to be *conditionally relevant* [21], meaning that although it is not probatively relevant by itself, it would be probatively relevant if proof could be introduced sufficient to prove it based on factual evidence. The meaning of the notion of conditional relevance is clarified by a remark in the FRE, Article 1, General Provisions, Rule 104 [7], under Preliminary Questions where it is called "Relevance That Depends on a Fact". The conditional relevance rule is quoted below.

---

[7]https://www.law.cornell.edu/rules/fre/rule_104

When the relevance of evidence depends on whether a fact exists, proof must be introduced sufficient to support a finding that the fact does exist. The court may admit the proposed evidence on the condition that the proof be introduced later.

In this sense of the term, P1 could be said to be conditionally relevant, meaning that its acceptability depends on whether the fact P10 exists, and provided proof can be introduced sufficient to support a finding that P10 is supported by enough evidence to warrant its acceptance. So now the question is whether P10 is supported by enough evidence to warrant its acceptance. To answer this question, we need to turn to Fig. 7.

Look along the bottom sequence of argumentation in Fig. 7 leading from P2

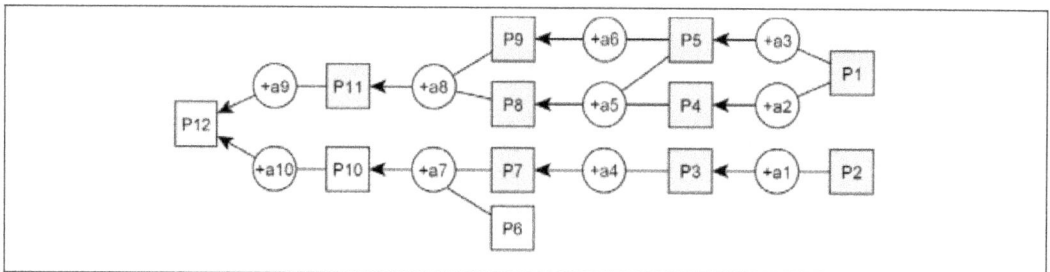

Figure 7: Failure of Relevance Shown in a Carneades Argument Map

towards P12. P2 is accepted, as shown by its appearance in a darkened (green) box. Therefore P3 is also accepted, since argument a1 has only one premise, which is accepted. And hence P7 is also accepted. But P10 is not accepted because one of the premises of a7, namely P6, is not accepted. Therefore argument a10 does to prove P12. Therefore P2, although accepted as factual, is not relevant to proving P10, which is in turn not relevant to proving proposition P12. But if P6 were to be accepted, P10 would be proved, and therefore P2 would also be relevant with respect to P12.

The structure of the abstract example displayed visually in Fig.'s 5, 6, and 7 suggests how distance in relevance is related to conditional relevance, how the two notions are connected, and how their relationship in argumentation can be represented using Carneades maps. But there are additional problems to be discussed about how this approach could be implemented.

9 Probative Weight

The question remains of how the notion of probative weight referred to in rule 403 of the Federal Rules of Evidence could be modeled in Carneades. Let us consider an abstract example (Fig. 8), where weights are attached to the arguments in the argument graph. In Fig. 8, five propositions are shown in rectangles with a

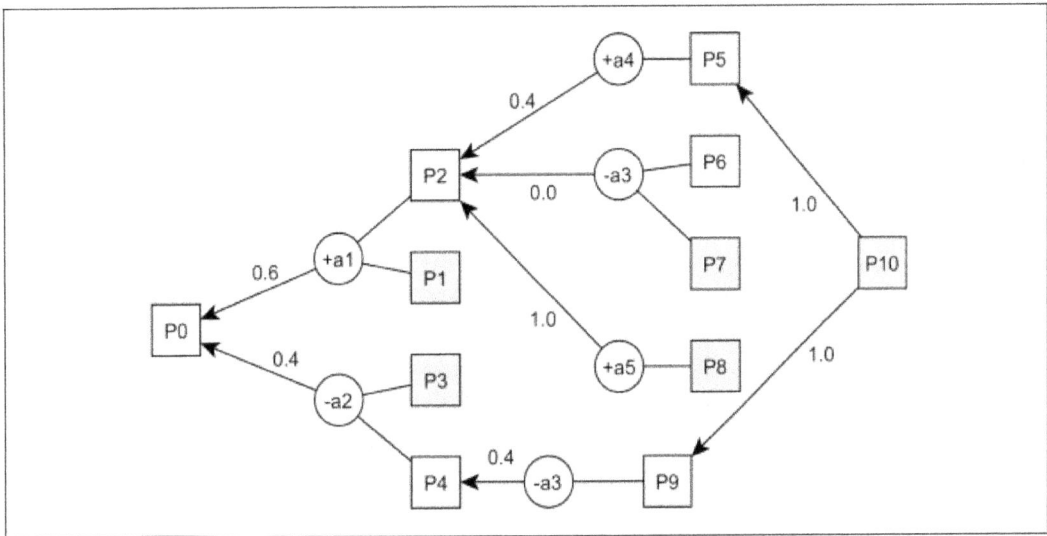

Figure 8: Evidential Graph 1 with Weights

green background indicating they have been accepted as evidential propositions by the audience. All the other propositions are âĂŸneither accepted nor rejected'. A number is attached to each argument representing its weight. In this instance, pro argument +a1 has a weight of 0.6, meaning that it has the potential to win out over con argument âĂŞa2, which only has a weight of 0.4. But how could the weights confirm or disconfirm this hypothesis?

To see how, let's look at Fig. 9, where inferences have been drawn from the premises accepted as evidential propositions. In Fig. 9 it is shown that all the premises of pro argument a1 have been proved, and all the premises of con argument a2 have also been proved. The weights break the deadlock, because the weight of pro argument a1 is greater than the weight of con argument a2.

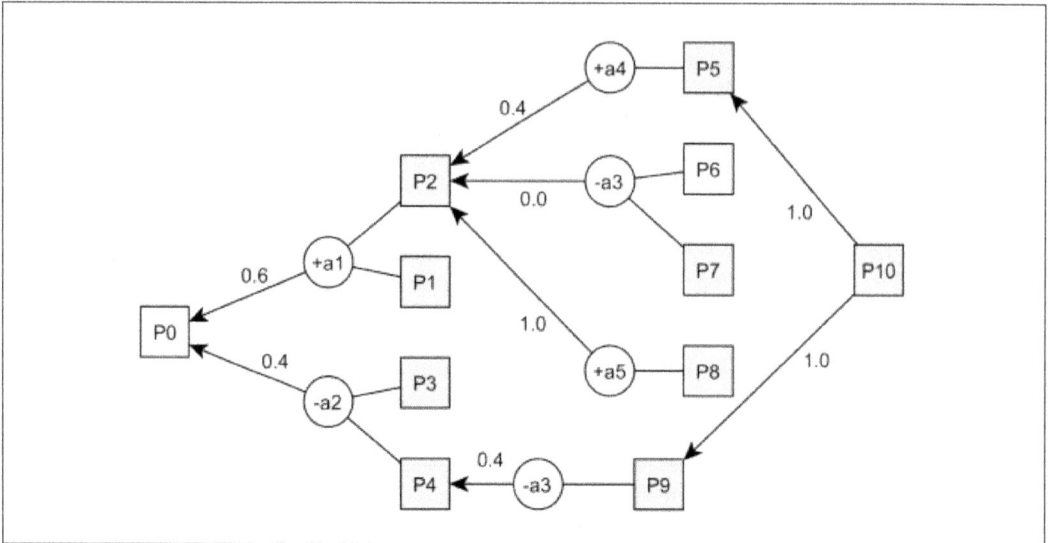

Figure 9: Evidential Graph 2 with Weights

10 Conclusions

This paper followed [27], [28], [22], [18], [11] and [13] in arguing that the way forward to define relevance is to combine argument maps, argumentation schemes, and the representation of implicit premises and conclusions on these argument maps. It has shown how, in modeling typical cases of legal argumentation where one argument can be indirectly relevant to another, this notion of argument distance can be conceptually sharpened by using a graph structure for argumentation. The specific argumentation system used for this purpose was Carneades. It used a series of argument maps of examples of legal argumentation showing how argument distance can be represented in cases where claims of relevance are refuted or supported using an evidential argument map. Such a graph can be used to show how one argument is indirectly relevant to the other by means of drawing an argument map depicting the series of intervening arguments that connect them. This tool can be applied to legal cases to represent an interpretation of a sequence of argumentation in a given natural language text going from an evidential bases to an ultimate conclusion (or not).

The chief concern of this paper has been probative relevance of arguments, and especially the notion of distance in a bipartite graph structure where each argument is connected to other arguments by means of premises and conclusions. Each example has been depicted in a framework of defeasible argumentation of this type in which there is a series of consecutive argument steps from the evidence to the facts at issue and from there to the ultimate *probandum* in a given case. Figure 4 represents the most general argumentation structure of this kind. In such a structure, an argument is probatively relevant to another argument if and only if the two are parts of a sequence of argumentation that (1) starts from some evidential propositions accepted by the factfinders and propagates forward through other arguments finally reaching the ultimate *probandum* proposition, and (2) each of the intervening arguments is defeasibly valid and has premises and a conclusion that is accepted by the audience. With such a sequence represented as a graph, it can be seen whether a given pair of propositions are directly relevant to each other or indirectly relevant, meaning that they are related to each other by more than one step. It can also be seen whether any item of evidence is relevant or not from the point of view of the argumentation in the trial as a whole. It can be used in this way to model inferential distance by using schemes, argument maps, and enthymemes.

Carneades was also used to show in Fig.'s 5, 6, and 7 to explain how distance in relevance is related to conditional relevance, how the two notions are connected, suggesting an approach to modeling relevance of evidence that could be implemented using computational argumentation tools. It needs to be emphasized however that what this paper has not done is to produce a formal and computational model of relevance in legal argumentation using Carneades. As shown in section 6, relevance in argumentation is a deeply pragmatic concept which varies depending on the dialectical setting of the argument in a given case at issue. In a legal trial in the common law system, the type of dialogue is that of a persuasion dialogue with three stages. At the opening stage, a burden of persuasion is set in place, along with the standard of proof required to fulfill the burden, and the elements of the ultimate *probandum* that need to be proved by the party having the burden of persuasion. Some evidence has already been gathered at the opening stage in order to show that a trial is appropriate based on the pleadings and the evidence known so far. Further evidence is collected during the argumentation stage, for example as witnesses testify and other admissible evidence and arguments are brought forth during Wigmore's five stages. The criterion of whether evidence is admissible is relevance. At the closing stage, a decision is made by the audience, the judge or jury, which side has won the case by a meta-dialogue which is used to examine the arguments on both sides, as happened in the bank robbery example.

The long and the short of this matter is that (1) relevance needs to be modeled in different ways in different legal and other formal dialogue settings, (2) this shows that it is an irreducibly pragmatic concept, and (3) for that reason we have a ways to go before being able towards modeling relevance using inferential distance, argumentation schemes and implicit premises. Still, it looks like we have some useful tools.

A possible one sidedness of this paper is that all of the examples treated have been instances where one argument or proposition is relevant to another. We have not tried to analyze any examples of irrelevance in argumentation. This latter is a topic of some interest in the field of logic, where in fact irrelevance has most prominently been treated under the heading of fallacies. It would be interesting also to find and analyze some examples of arguments in a legal setting that can be shown to be irrelevant by the methods set out in this paper.

References

[1] Atkinson, K., Bench-Capon, T. J. M. and Walton, D.. Distinctive Features of Persuasion and Deliberation Dialogues, *Argument and Computation*, 4(2),, 105-127 2013.

[2] Bex, F. J. *Arguments, Stories and Criminal Evidence.* Dordrecht: Springer, 2011.

[3] Dung, P. M. On the Acceptability of Arguments and its Fundamental Role in Nonmonotonic Reasoning, Logic Programming and n-person Games. *Artificial Intelligence*, 77(2), 321âĂŞ357, 1995.

[4] Epstein, R. L. Relatedness and Implication, *Philosophical Studies*, 36(2), 137-173, 1979.

[5] Epstein, R. L. *The Semantic Foundations of Logic, Vol. 1, Propositional Logics*, Dordrecht: Kluwer, 1990.

[6] Gordon, T. F. The Carneades Argumentation Support System, Dialectics, *Dialogue and Argumentation*, C. Reed and C. W. Tindale (eds.). London: College Publications, 145-156, 2010.

[7] Gordon, T. F. and Walton, D. Formalizing Balancing Arguments. *Proceedings of the 2016 conference on Computational Models of Argument (COMMA 2016)*, IOS Press, 327-338, 2016

[8] Gordon, T. F., Friedrich, H. and Walton, D. Representing Argumentation Schemes with Constraint Handling Rules, *Argument & Computation*, 9(2), 91-119, 2018.

[9] Hohmann, H). Dynamics of Stasis: Classical Rhetorical Theory and Modern Legal Argumentation, *American Journal of Jurisprudence*, 34, 171âĂŞ197, 1989.

[10] Kennedy, G. *The Art of Persuasion in Ancient Greece.* Princeton: Princeton University Press, 1963.

[11] Kok, E. M., Meyer, J-J., Prakken, H. and Vreeswijk, G. A Formal Argumentation Framework for Deliberation Dialogues. In *Argumentation in Multi-Agent Systems,* ed. McBurney, P., Rahwan, I. and Parsons, S. Berlin: Springer, 31âĂŞ48, 2011.

[12] Lawrence, J. and Reed, C. Combining Argument Mining Techniques, *Proceedings of the 2nd Workshop on Argument Mining.* Copyright: Association of Computational Linguistics, 127- 136. http://www.arg.dundee.ac.uk/people/chris/publications/2015/ArgMining2015.pdf, 2015.

[13] Macagno, F. Assessing Relevance, Lingua, JulyâĂŞAugust, 2018, 42-64. https://www.researchgate.net/publication/324755120_ Assessing_ relevance, 2018

[14] Mauet, T. A. *Trials: Strategy, Skills, and the New Powers of Persuasion.* New York: Aspen Publishers, 2005.

[15] McBurney, P., Hitchcock, D. and Parsons, S.. The Eightfold Way of Deliberation Dialogue, *International Journal of Intelligent Systems*, 22(1), 95âĂŞ132, 2007.

[16] Michael, J. and Adler, M. The Trial of an Issue of Fact: I, *Columbia Law Review*, 34, 1224-1306, 1934.

[17] Park, R. C., Leonard, D. P. and Goldberg, S. H. *Evidence Law,* St. Paul, Minnesota: West Group, 1998.

[18] Prakken, H. Models of Persuasion Dialogue. *Argumentation in Artificial Intelligence*, ed. I. Rahwan and G. R. Simari, 281âĂŞ300. Dordrecht: Springer, 2009.

[19] Strong, J. W. (ed.). *McCormick on Evidence*, 4th ed. St. Paul, Minnesota: West Publishing Co. 1992.

[20] Twining, W. *Theories of Evidence: Bentham and Wigmore.* London: Weidenfeld and Nicolson, 1985.

[21] Walton, D. Argumentation Schemes: The Basis of Conditional Relevance, *Michigan State Law Review*, 4 (winter), 1205-1242, 2003.

[22] Walton, D. *Relevance in Argumentation.* Mahwah, N. J.: Erlbaum, 2004.

[23] Walton, D. and Krabbe, E. C. W. *Commitment in Dialogue*: Albany: SUNY press. 1995

[24] Walton, D., Reed, C. and Macagno. *Argumentation Schemes.* Cambridge: Cambridge University Press, 2008.

[25] Walton, D. and Gordon, T. F. Formalizing Informal Logic, *Informal Logic*, 35(4), 2015.

[26] Walton, D. and Macagno, F. Profiles of Dialogue for Relevance, *Informal Logic*, 36(4), 523-562, 2016.

[27] Wigmore, J. H. *The Principles of Judicial Proof.* Boston: Little, Brown and Company, (second edition), 1931.

[28] Wigmore, J. H. *Evidence in Trials at Common Law, vol. 1a*, ed. Peter Tillers, Boston, Little, Brown and Company, 1983.

Received 20 May 2019

www.ingramcontent.com/pod-product-compliance
Lightning Source LLC
Chambersburg PA
CBHW081339090426
42737CB00017B/3208